RESEARCH BUILDING
PLANNING AND DESIGN

Edited by Neil Appleton

DESIGN MEDIA PUBLISHING LIMITED

FOREWORD

The Salk Institute, designed by Louis Kahn's practice in the early 1960's, stands today as one of the most revered and productive biomedical research building in the world. It has housed a number of Nobel laureates, and the institute is consistently ranked as a world leader in the fields of neuroscience and plant biology. What makes it such a great institute, some 50 years since its foundation? Certainly the reasons are many, but it's worth considering one of its core design concepts, the idea of an architecture that clearly facilitates both team collaboration and individual contemplation within the new facility.

The large flexible labs were conceived of as the site for dynamic 'bench top' research, whilst the monastic office towers, with famous views to the Pacific Ocean, form the 'professor's retreat'. At its core the design idea held the scientist's ability to discover through discourse (experimental) and introspection (imagination) as crucial, supposing that the combination of these 'thinking' modes would yield significant results. Not to mention the great architectural realisation of the conceptual diagram, surrounded by charming environs.

Designing flexible and adaptive laboratory spaces are a given, necessary technical skill requisite of any architect in this field. Research teams will grow and contract in sync with funding cycles and the success of the research, and new super sensitive optical instruments and high throughput equipment will move the research staff's focus from time spent at the wet bench to time spent in dry analytic labs. The days of each researcher owning a lab bench are numbered, with more 'hot benches' likely to become the norm. In concert bio-containment regimes will continue to move from the macroscopic to the microscopic.

But none of these technical demands are, in our view, the key to the design of great laboratory research buildings. The best research building designs locate the researcher as a creative agent at the centre of the process. The building is not a machine in which the scientist is but a tooth in cog. The building is a cultural construct that supports collaborative tension and imaginative play.

With the thought that the next wave of great research findings will come from cross-disciplinary interaction and a greater emphasis on translational research (involving clinicians, research scientist and industry technologists for instance), the value of diverse modes of interaction between researchers is now in the foreground of the design process. Designing for 'soft' research space that encourages happenstance meetings and personal delight is critical to the imaginative creative process. At a fundamental level researchers are trying to see the unseeable, or think the unthinkable, making new things, relevant and pertinent to the human condition. A factory or banal office space may not be the best environment for this process.

In the global research marketplace, large 'new' research institutes have become critical mass 'attractors' for the very best scientists. In Australia in the last 10 years the federal government has invested in building the 'clever country' through an ambitious research infrastructure-funding project, in an attempt to reverse the 'brain drain' to off shore institutes. 500 or so researchers in one building have become a relative norm. These are recognised as places for social interaction as much as professional collaboration.

How to make these buildings attractive beyond just a well designed laboratory and well stocked equipment barn? Researchers demand an environmentally sustainable workplace, with abundant fresh air and natural light within a low energy use building. The new research institute has homely social 'rooms' like a communal kitchen, a collaboration lounge, a journals library and a games room. Artwork adorns the walls. Green roofs function as living laboratories. Open interconnected stairs encourages researchers to zip between levels. Hi-resolution 'telepresence' connectivity to international colleagues is commonplace. Nooks and crannies are fashioned for write-up and quiet contemplation. Team meetings and seminars are held locally. Teaching and mentoring spaces are interconnected, one easy step from the research laboratory. Industry partners are encouraged, welcomed and accommodated throughout the building. Exhibitions are held in the large engagement entry hall, building public awareness and engagement in the research endeavor. Barista made coffee ('Melbourne Style') and a topical bookshop are essential in the mix.

Finally it is our view that, in the design of new research buildings as significant 'attractors', architecture can be part of the exciting story of contemporary research, expressive of an experimental world, dynamic and discursive... engaged in the world of research ideas through metaphor and materiality. An architectural identity can resonate beyond a mere commercial type, to become a great research 'institution', recognised and respected both locally and globally. The Salk Institute is one great enduring example.

Having designed many significant research buildings in the last 10 years, our practice has had the pleasure in working collaboratively with a broad cohort of research scientists to create compelling and expressive architecture. The scientist's creative capacity, analytical ability, and 3D visualisation skill make them a great partner to join in a narrative design journey with the architect.

Neil Appleton
Director, Lyons
Melbourne
2013

CONTENTS

What's the Research Building

As a materialised space for scientific research, the research building is a socialised architecture designed for teamwork research. The laboratory is the core of the building because the development of contemporary science and technology is closely related to scientific experiments. Research depends on the application of the scientific method, a harnessing of curiosity. The research provides scientific information and theories for the explanation of the nature and the properties of the world. Researches can be divided into different classifications according to their academic and application disciplines. Laboratory buildings can also be divided into different classifications according to their specialisations (research, education and manufacture) and scientific disciplines (chemistry, physics, biology, medicine, etc.). Therefore, besides reflecting the characteristics of the discipline which the laboratory belongs to, the seemingly simple working tables and experiment operating programme must need supports from professional equipments, piping system and electric system. In other words, the emergence of research buildings is the inevitable demand of the development of scientific researches. Meanwhile, the development of research buildings is the reflection of the continuously improved demands of scientific researches.

Generally, research is understood to follow a certain structural process, observations, hypothesis, gathering and analysis of data, and finally conclusion. According to the process, research is totally teamwork. No one can do all the procedures, so every scientist should be responsible for one of those procedures, and then they are in collaboration with others to conclude. Members of the team should not only spend much time in their own laboratory alone, but also they should talk to their colleagues and discuss their new hypothesis and question about the research. Therefore, the research building should provide an open seminar and meeting room to accelerate the process of research. It should provide ample spaces for discussion and conversation. Scientists and their technical assistants spare no efforts to exchange their new findings, and make creative definition of the topic. The whole team becomes a well-functioning body. All the smart "brains" put forward the innovative hypothesis, and their "eyes", "arms" observe and test the hypothesis, and again and again revise the analysis.

Individually, teams are composed of scientists and technical assistants. The fruit of the whole team relies on every member of the team. Every research may involve sitting in front of computer screens, and observe small objects under the microscope for many an hour around the clock. Hence, individual office cells may provide relatively quiet surroundings for researchers to focus on their own observations and calculations. Observation and calculation are of great importance, for researchers should take notes of the every step of experimental process, and instantly get down to the date analysis. Researchers should not miss or ignore every detailed change, or those neglected details may probably lead to the failure of the experiment. Therefore, a comfortable laboratory environment will stimulate creation for researchers with intensive working. This kind of environment is a resource, for a pleasant office atmosphere and appropriately private working environment will be helpful for the researchers to do better scientific experiments.

The research building should also provide an environment of warmth and security. The building designer should not only take the well-equipped facilities into consideration, but also they may create a comfortable ambience for diligent and industrial scientists to enhance the quality of their work life. On the one hand, many scientists may burn the midnight oil, and even sleep in their office. So providing nutritious breakfast and a cup of sweet coffee may make scientist feel themselves at home. Providing food is very simple, but it will encourage researchers to spend more time in researching, and pursue their goals in academy with more diligence. On the other hand, apart from laborious research, scientist should spend much time in exercising and working out considering their health. Researchers may sit all day around in front of their computer. In this way, scientists may feel tired of their work day by day. The research building should provide a range of sports facilities.

The Stages of Development of the Research Building

New life styles will introduce new types of architecture. As the architectural space is where people work and live, the types of architecture will change according to mankind's modes of production and life. The scientific research can be divided into three stages: unified ancient philosophic science, scientific classifications from the Renaissance to the late 18th century and comprehensive science from 19th century to 20th century.

1. The Laboratory Space in the Budding Stage of Science

In this stage, science was not an independent discipline but included in philosophy. At that time, many philosophers were working on scientific researches and experiments. With the influence of religion, the science was mostly focused on alchemy and astrology. Besides, limited by the scientific and technological level, the laboratories were simple and crude, usually located in residential buildings. Researchers worked on the initial experiments in these simple laboratory spaces, which gradually developed into normal laboratory buildings for researches. (See Figure 1)

2. Laboratory Buildings with Clear Discipline Classification

In the early stage of the development of scientific research, research was still an individual activity. Therefore, there was no specific buildings for researches. Since the Renaissance, science became the discussion of a certain knowledge field because different fields need different experimental means and methods. After that, with the classification of natural science disciplines, the laboratory buildings were classified according to their specialisations (research, education and manufacture) and scientific disciplines (chemistry, physics, biology, medicine, etc.). Also, with the complication of experimental procedure, there emerged some specific discipline laboratory units consisting of several laboratories. They formed closed laboratory buildings with clear discipline classification. (See Figure 2–4)

Figure 1. A Medieval laboratory – from Baidu

3. Modern Scientific Laboratory Buildings with Diverse Spaces

Flexible Laboratory Spaces

(1)The comparably large open space is divided into several small units with different experimental activities and the staff are separated too.

(2)The connecting passages between spaces weave the separate spaces together and the organisation of transportation lines should provide convenience for the experiments.

Modern Research Buildings

With the social changes and the scientific and technological progresses, the development of science tends to be integrated and intensified, which needs efficient and large-scale organisation and coordination to conduct a series of scientific researches. Simple laboratory spaces can no longer satisfy the requirements of the development of contemporary science and the multi-

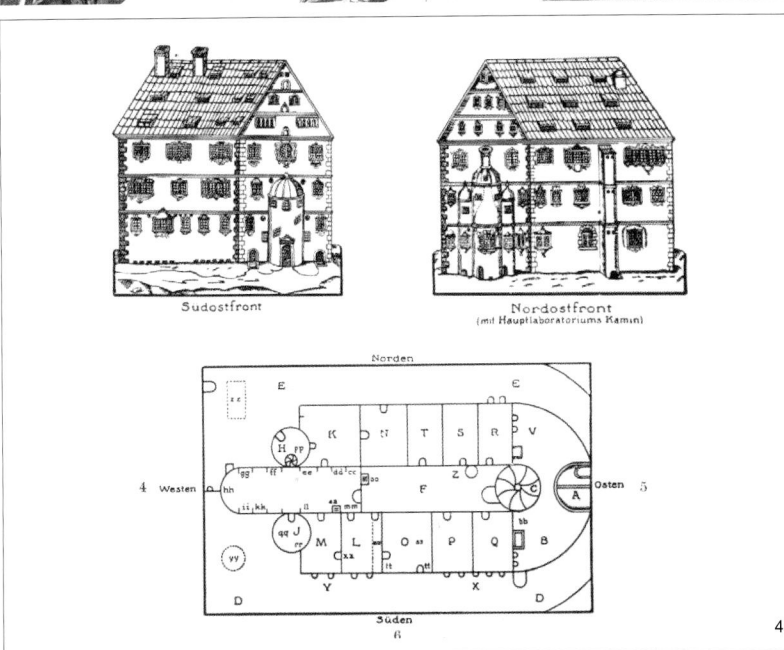

Figure 2. An 18th century chemistry laboratory from Diderot's Encyclopaedia

Figure 3. Andreas Libavius (1555 – July 25, 1616) was a German doctor and chemist.

Figure 4. This idealised design for a chemical institute in Andreas Libavius, Alchymia, 1606. It contains a main laboratory with furnaces for water-baths, ash-baths, and steam-baths; distillation apparatus for upward and downward distillation, with and without cooling; sublimation apparatus, fireplace; reverberatory furnace and large bellows. The analytical laboratory contains assay furnaces and analytical balances, some in cases. The private laboratory contains a philosopher's furnace. There is also in the institute a preparation room with press, a pharmacy, a crystallisation room, etc. The laboratory has water laid on, and in the open air there are facilities for making alum and vitriol, and a saltpetre plantation.

functional complex is becoming more favourable to researchers.

In the filed of biology and medicine, the research requires experimental animals to grow up in a certain condition which is different from normal atmospheric environment. Thus the biological cleaning technology emerges. The animal labotory adopts completely closed environment with manual control, no-barrier large spatial layout and changing cage locations according to different breeding requirements. In order to meet requirements of illumination and ventilation, the researchers not only control the storey height of the research building, but also take full advantage of the roof and illumination intensity difference to conduct experiments. (See Figure 5)

4. The Diversified Development of Research and Laboratory Buildings

(1) From Single Building to Multi-wing Complex

Ever since Andreas Libavius designed the first laboratory building all over the world in the 17th century, various types of research buildings and composite design methods have emerged. The scale and scope of scientific research in the 17th century decided that the single building would meet the demand. Individual buildings separated different disciplines spatially and cut off their relationship. With the development of industrial manufacturing and the expansion of education and research institute, single buildings can no longer meet the development of modern science's requirements. For example, there is a design method to separate the research rooms from the main building, so that the research rooms are both independent from the laboratories and connected to them, which forms a wing composite plan, such as the Bell Laboratory in America. (See Figure 6)

(2) The Diversification and Standardisation of the Floor Plan

In 1960s, people developed an architectural form which took a standard-size space as the traffic junction and could be developed horizontally. The single volume can be flexibly divided into large areas according to the requirements of researches. The flexibility of the building was thus achieved through large areas and the junction.

The spatial forms of research buildings have gradually got rid of scattered layouts and moved towards intensified integration. They present an open status in terms of time and space. And with the reasonable sharing of physical resources, it forms an integrated mode of scientific research, study and life. (See Figure 7)

Figure 5. The iconic laboratory. Heinrich Wieland in the Baeyer Chemistry Laboratory, University of Munich, circa 1925. Edgar Fahs Smith Collection P/L 112.23 M. Courtesy of Rare Book and Manuscript Library, University of Pennsylvania.

Figure 6. Bell Labs' headquarters is in the USA was designed by Eero Saarinen. The 2,000,000-square-foot building was constructed between 1957 to 1962, which could contain 6,000 staff.
Figure 7. The length of the perimeter corridors were kind of daunting but really amazing with the light.

What makes research buildings characterise? What makes research buildings differ from other buildings? What makes research buildings become research buildings? You may find these answers in the following part.

Generally, many factors make great influence on the design and construction of the building. In addition, research buildings, as the representative of knowledge-intensive occupations in our industrial society, demand more special designs in power-saving and environmental protection.

Take Great Advantage of Significant Local Elements

Research buildings can take great advantage of local geographical factors. When it comes to the weather, we cannot but think of the temperature. Many research buildings are located in the places where it is very hot and with abundant sunshine. As the saying goes, each coin has two sides. On the one hand, the design team can apply the play of the light and shade to botanical research. On the other hand, in response the extreme temperature and intense heat of the sun, designers have to try new technical ways to reduce the indoor temperature. Apart from temperature, the humid climate can also be taken advantage of. Although highly humid climate may result in the uncomfortable feelings, designers are so intellectual that these problems can be solved. Moreover, the canted roof directs rainwater into a large underground cistern that retains the water on site to help recharge the ground water, and enrich the soil for the surrounding landscape. Therefore, scientists can make most of the water to make experiment. In addition, it is vital for designers to take terrain factors into consideration. Many of the research buildings are set into the contours of gently sloping site. Hence, designers must have a second thought of the foundation. Certainly, most of research buildings are constructed on the flat plain.

Diversity of Inner Functional Space

Research building has a wide variety of facilities, including labs, conference and meeting rooms, public spaces, exhibition space, restaurants, administrative services and offices. According to the kind of research, the research building takes on different layouts. In general, there will be parking services in the basement. The ground floor, taking up the whole plot of land, holds common spaces and the entrance hall. The research facilities are developed on other floors. Moreover, the building's green roof can be laid with local moss, serving as a filtering mechanism for rainwater and habitat for birds and native insects. The green roof can serve as a research centre for botany students, and the plant and soil can be used in research. A green roof can also save energy because they increase the insulation of the roof and it tends to last longer than standard roofing. Also, green roof improves air quality around the building, and slow down the water of storm during rainy seasons.

Pay Great Attention to Open Elements

Another design element we should pay special attention to is open spaces. Between two buildings, the open route is the interface with the adjacent building, nurturing and supporting communication between the two facilities whilst allowing staff to meet and chat informally within the circulation spaces. In the same building, but on different floors, from one space open to the public as well as the entrance to the laboratory floors above are accessible. Even in the same floor, designers can deliver an open and adaptive laboratory/office environment. Open office and meeting room layouts without walls, or columns smooth the exchange of knowledge and improve collaboration. Research Buildings should not only meet current requirements, but fit future requirements to realise a flexible interior spatial design. (See Figure 1)

Use New Materials and Hi-Tech

Architects also apply the new material and hi-tech to the construction of research building, and they try every means to reduce the amount of energy required, and operating costs. They use the façade structural and technical measures to achieve power-saving purpose, and they may put solar photovoltaic panel on the top of the building to save the energy. For

Figure 1. An open and adaptive laboratory/office environment

example, they may use solar water heating panels as a system that uses the sun's energy rather than electricity or gas to heat water. A solar water heater uses glased collectors that are roof-mounted and connected to a preheat storage tank. Fluid is pumped to the collector where it is heated by the sun, and returned to a heat exchanger where heat from the fluid is used to heat the water in a preheat storage tank. The system can not only provide a large proportion of a building's water heating requirements, but also can operate at minimal cost. The designers can also choose high performance glass and sunscreens to manage sunlight and heat gain. A typical laboratory building's energy consumption is three times as much as that for an office building. Many laboratories need a large amount of electric power and continuous power supply. In a world with energy emergency, a laboratory building should not only meet the requirements of safety and health, but also take sustainable design into consideration.

Research is a competition in nowadays industrial society. Therefore, the research building is also part of the competition. Sometimes the research building may appropriately symbolise the company's image. Science and technology constitute a primary productive force, so is the research building.

The expansion of scientific development and the population working on science, the relationship between science and manufacturing, the combination and interaction between research buildings, the rapid development of scientific research and the requirement for laboratory buildings to change rapidly according to the demand of research process...these factors all put forward more requirements for the design and construction of research buildings.

Demonstration on the Characteristics of Modern Science Buildings

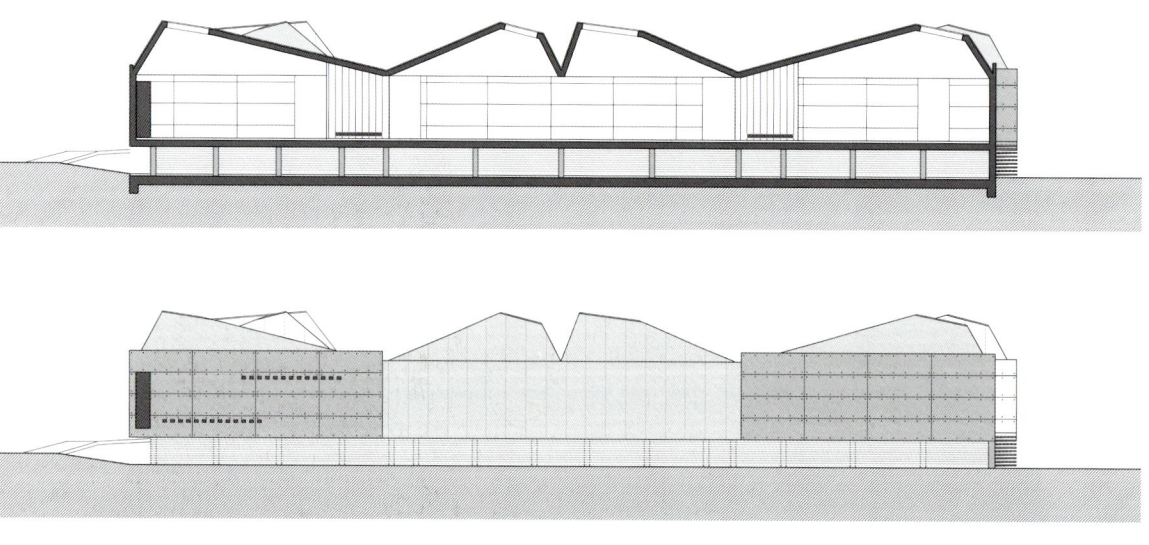

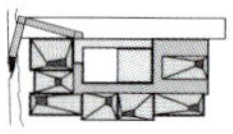

Section and South Elevation

1

Efficient use of advantages of the location
| Centre for the Interpretation and Research of Rivers: Órbigo

Location:
Benavente, Zamora, Spain
Architect:
JOSÉ JUAN BARBA Architect
Construction Area:
900 m²
Completion Date:
2009
Photographer:
Ignacio Bisbal Grandal, José Juan Barba

- Programmatically the project is outlined as the grouping of five modules around a courtyard on two levels.
- The design aim is to maximise the combination between the building and its surroundings and reflect the relationship between artificiality and nature.
- minimal impact on the surrounding nature
- intervening in a semi-natural environment with criteria of passive sustainability
- using materials whose ageing process allows the building to converse with the changes in tone, colour and light of the surroundings.

- The site's influences to the architectural form: Its condition of a flood plain, situated in a fluvial valley defines the solution proposed from its beginnings. Therefore the building is elevated above the natural terrain by means of a system of piers. Access is gained by means of a ramp, which serves for penetration.
- The exit is equipped with a roofless glass corridor. The design of two glass walls and non-existent roof integrates the visitors into the nature and provides an experience of walking along the river.

1. South elevation; the whole project, presented as a single architectural element, develops its five thematic areas as a single room, its routes embracing the two courtyards, which represent two opposites, artificiality and nature.
2. Night view of south elevation
3. When we are in the exterior, the evidence of the vertical component is intensified by contrast; there is no roof and the sides are almost incorporeal.
4. Northeast view of the building
5. The exit after the tour is along a roofless corridor with glass walls. This is the only time one loses the direct views of the courtyards although their presence is still felt. The two glass walls and the absence of a roof attempt to give the visitor the feeling of going along inside a river.
6. East access

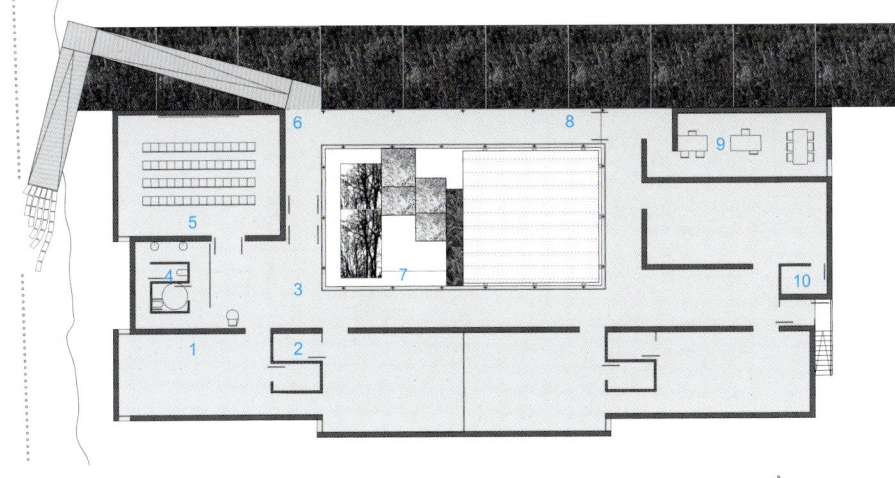

10

7. Access area
8. Access to the 1st floor exhibition room
9. The 1st floor exhibition room at night
10. In the project the creation of places is based on
a complete understanding of space and the influence
of its different escapes. The vertical component of
this interior space is reflected in the domes and its
escapes through the skylights.

Ground Floor Plan

1. Hall
2. Storage
3. Foyer
4. Washroom
5. Screening room & projection hall

6. Access
7. Garden
8. Gangway
9. Administration area
10. Installation

11. The tour of the Centre begins with a projection room where the visitor receives his first immersion in information.

12. Bathroom

13. A new landscape or the extension of the surrounding landscape is contemplated in the interior of the project.

13

TIPS:

Ignasi de Sola Morales said "...architecture is an act of violence, because it changes the nature of the materials, it uses the place where it is situated." – it transforms them, it models them and therefore is violent towards the space on creating something new which did not exist previously. A violence understood as the change which occurs in living beings when they grow, walk and live.

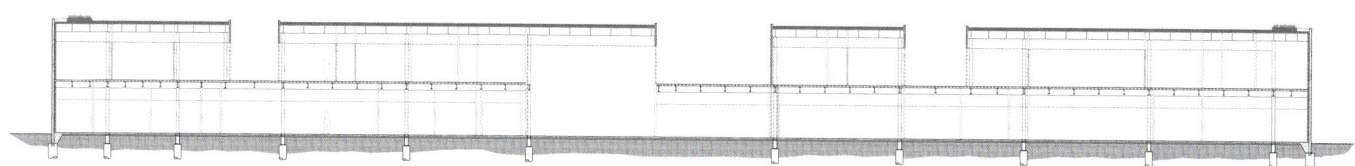

East-West Section through Think Block

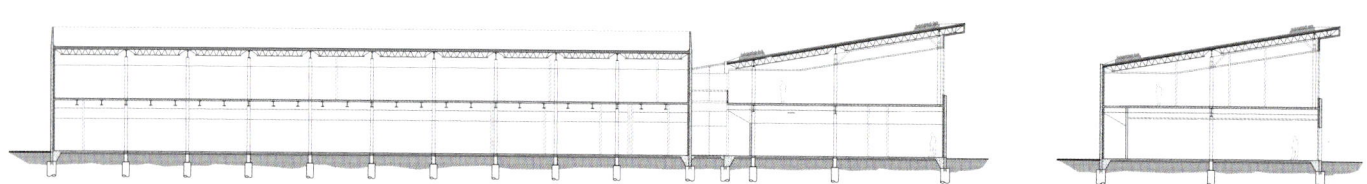

North-South Section through Archive Block

North-South Section through Think Block

Unique design of façade tells a great deal about the local natural features
| Botanical Research Institute of Texas

Location:
Fort Worth, Texas, USA
Architect:
H3 Hardy Collaboration Architecture
Construction Area:
6,500 m²
Completion Date:
2011
Photographer:
Chris Cooper

▪ The textures and grains of the concrete wall will change according to the seasonal sunlight and appear different in days and nights.
Detailed Description: The building's exterior character acknowledges the intense climate of Fort Worth in site placement, materials, and façade design. On the concrete façades, overlapping vines of different patterns of foliage, texture, and colour respond to shifting sun angles during the day and changing seasons throughout the year.
▪ The location and physical environment:
Fort Worth: Fort Worth is located in North Texas and in the South portion of the United States. It

has a humid subtropical climate according to the Köppen climate classification system.
▪ In the design of archive building, the designer chooses innovative design to maximally control the temperature and humidity, providing a better environment for the saving of documents.
▪ Interior design points: Floor to ceiling glass on the north façade fills spaces with generous natural light.
▪ Energy Saving and Sustainable Design:
LEED-NC Platinum certification
Bamboo ceiling panels
Dramatic sinker cypress feature wall

1. The design clearly organises a variety of separate yet interconnected functions by dividing the building into two structures: the "Think Block", housing the administration and research offices, education department, exhibit area, and public spaces, and the "Archive Block", housing the herbarium and library.

2. The long two-storey rectangular structure of precast concrete panels is punctuated with glass – broad expanses on the north side to bring in plentiful light and smaller openings on the south.

3. Because of the delicate nature of the specimens, the Archive Block is nearly windowless – a solid box of tilt-up concrete panels to provide maximum temperature and humidity controls.

4. The sloping living roof of the Think Block is covered with informal patterns of regional sedums and grasses, providing a niche for preserving the beautiful Fort Worth Prairie in a new form for the future.

5. These lush, colourful plantings lead visitors from the parking lot to the organic trellis work canopy that shades their path into the building.

9

6. Bamboo ceiling panels and a dramatic sinker cypress feature wall in the lobby highlight a sophisticated panoply of materials that are healthy, sustainable, handsome, and durable.
7. The office space is cantilevered out from the north façade, providing further visual connection to the greenery outside.
8. Floor to ceiling glass on the north façade fills spaces with generous natural light.
9.10. The interior character of the building offering an environment where research, informed study and collaboration can take place in a collegial, inspiring atmosphere.

10

11. Corridor of the education centre
12. The Archive Block houses the extensive collection of botanical specimens in the Herbarium, a two-storey climate controlled storage hall, together with a distinguished book collection found in the library stacks.
13. Sky-lit bridge space that connects the two Blocks and provides orienting views to the outside

Ground Foor Plan

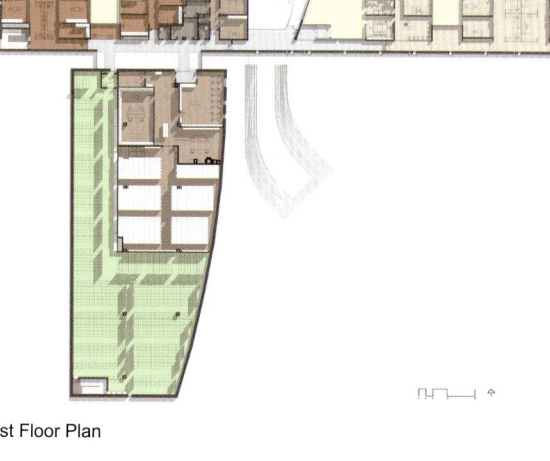

1st Floor Plan

Administration

Advancement

Building Services

Collections Management

Education

Herbarium

Library

Outdoor

Press

Public spac

Research

TIPS:

1. "Systematics"
Biological systematics is the study of the diversification of living forms, both past and present, and the relationships among living things through time.

2. Bamboo ceiling panels
In its natural form, bamboo as a construction material which is traditionally associated with the cultures of South Asia, East Asia and the South Pacific, to some extent in Central and South America and by extension in the aesthetic of Tiki culture. In China and India, bamboo was used to hold up simple suspension bridges, either by making cables of split bamboo or twisting whole culms of sufficiently pliable bamboo together.

Bamboo intended for use in construction should be treated to resist insects and rot. The most common solution for this purpose is a mixture of borax and boric acid.

Bamboo has been used as reinforcement for concrete in those areas where it is plentiful, though dispute exists over its effectiveness in the various studies done on the subject. Bamboo does have the necessary strength to fulfill this function, but untreated bamboo will swell from the absorption of water from the concrete, causing it to crack.

1

2

3

The climate features are utilised to reduce energy use and the exterior materials have little need of future maintenance

| Paradise Valley Community College Life Science Building

Location:

Phoenix, Arizona, USA

Architect:

Marlene Imirzian & Associates LlC

Construction Area:

3,252 m²

Completion Date:

2009

Photographer:

Bill Timmerman/ Timmerman Photography Inc.

▪ The challenges of local natural conditions: The PVCC Life Science Building is a fusion of the context of sustainable and responsive desert architecture. It is of great necessity that the new Life Sciences building respond to the extreme temperatures and intense heat of the desert sun.

▪ The major exterior materials, masonry and copper, have no finish or coatings and have need of little future maintenance. The locally sourced copper (a major Arizona export) kept environmental and economic impact of

material imports low. Interior primary flooring is finished concrete and exposed masonry.

▪ The large "porch" roof shades the glazing and exterior spaces below reducing the temperature and making the exterior spaces usable for most of the year.

▪ Sustainability Features:

>>The building orientation reduces heat gain during the summer.

>>The white roof with high albedo reduces the heat island effect.

>>The canted roof directs rainwater into a

4

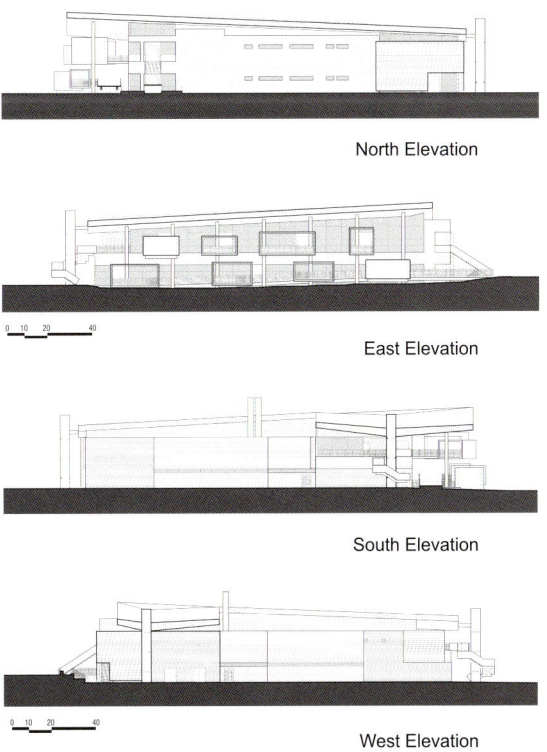

large underground cistern that retains the water on site to help recharge the ground water, and enrich the soil for the surrounding landscape.

>>Daylighting is provided in all occupied spaces, including laboratories, laboratory preparation, and classroom spaces, less energy is required for lighting.

>>The exterior circulation, stairs and collaboration space require no conditioning, reducing energy consumption. These elements take advantage of the eight months of good weather in the Sonoran Desert.

>>Digital control systems ensure automated operation of laboratory and building systems, optimising system efficiency.

>>Landscaping around the building create cool microclimates, used as learning and studying areas. The plants further help soil stability on site, control surface runoff, and provide a buffer between sunlight and building.

North Elevation

East Elevation

South Elevation

West Elevation

6

1. The façade & pods are shaded by a great roof overhang. The raised campus walk acts as a bridge to allowing the desert landscape to extend uninterrupted.

2. A learning landscape features Sonoran Desert plant life and an outdoor classroom.

3. Campus walk and collaboration pods and future campus green intertwine to promote informal learning.

4. The outdoor space becomes a great interaction zone.

5. The exterior circulation, stairs and collaboration space require no conditioning, reducing energy consumption. These elements take advantage of the eight months of good weather in the Sonoran Desert.

6. Bridged Connections push collaboration pods out into campus, and encourage informal gatherings and frame views to campus green and hills beyond.

7. West entry and stairs.

8. First floor interior circulation with connection to collaboration pods

9. Biology Lab. Daylighting is provided in all occupied spaces, including laboratories, laboratory preparation, and classroom spaces, so less energy is required for lighting.

10. Lab preparation space

TIPS:

1. Awards Name:

American Institute of Architects Arizona Design Award, Merit

Valley Forward Crescordia Award for Environmental Excellence

2. Phoenix, Arizona, USA

Phoenix has a subtropical desert climate (Köppen: BWh), typical of the Sonoran Desert in which it lies. Phoenix has extremely hot summers and warm winters. The average summer high temperatures are some of the hottest of any major city in the United States, and approach those of cities such as Riyadh and Baghdad. The temperature reaches and exceeds 100°F (38°C), on average for 110 days of the year, including most days from late May through to early September. Highs top 110 °F (43 °C) an average of 18 days during the year. On June 26, 1990, the temperature reached an all-time recorded high of 122 °F (50 °C).

3. Recycling of Copper:

Copper, like aluminum, is 100% recyclable without any loss of quality whether in a raw state or contained in a manufactured product. In volume, copper is the third most recycled metal after iron and aluminium. It is estimated that 80% of the copper ever mined is still in use today. http://en.wikipedia.org/wiki/Copper - cite_note-26 According to the International Resource Panel's Metal Stocks in Society report, the global per capita stock of Copper in use in society is 35kg~55 kg. Much of this is in more-developed countries (140kg~300kg per capita) rather than less-developed countries (30kg~40kg per capita).

The process of recycling copper follows roughly the same steps as is used to extract copper, but requires fewer steps. High purity scrap copper is melted in a furnace and then reduced and cast into billets and ingots; lower purity scrap is refined by electroplating in a bath of sulfuric acid.

1St Floor Plan

1. Laboratory	5. Circulation
2. Classroom	6. Outdoor collaboration space
3. Office	7. Rest room
4. Lab preparation	8. Office support

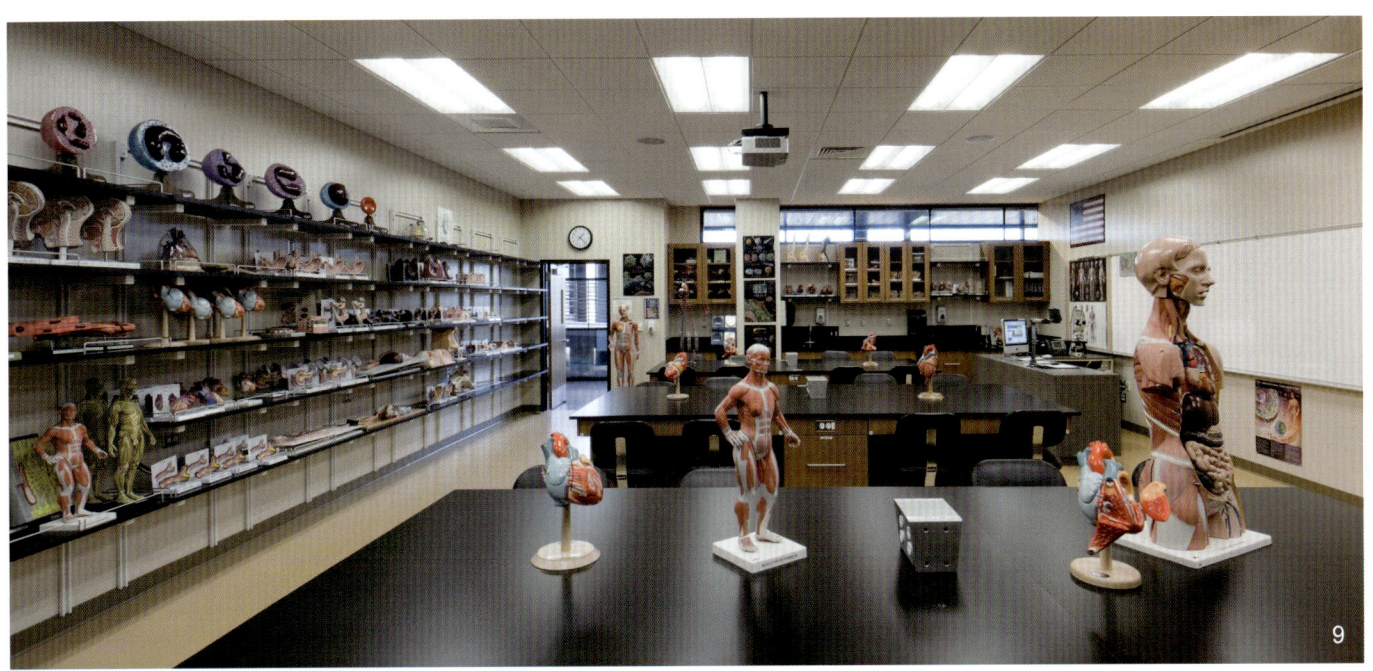

9

10

2

For the clients the design of atrium and greenhouse provides an optimal collaboration and communication environment

| Koppert Biological Systems

Location:
Veilingweg, the Netherlands
Architect:
HABEON ARCHITECTEN
Construction Area:
5,400 m²
Completion Date:
2010
Photographer:
Habeon, Koppert & Dimmy Olijerhoek

▪ The extending large undulant roof is an apparent design feature.
▪ In material and colour palette, the designer tries to convey a balanced and peaceful feeling, based on which, the functions of different departments are clearly relected.
▪ The designer uses open layout in interior design, while the design of atrium and greenhouse also provides an optimal collaboration and communication environment.

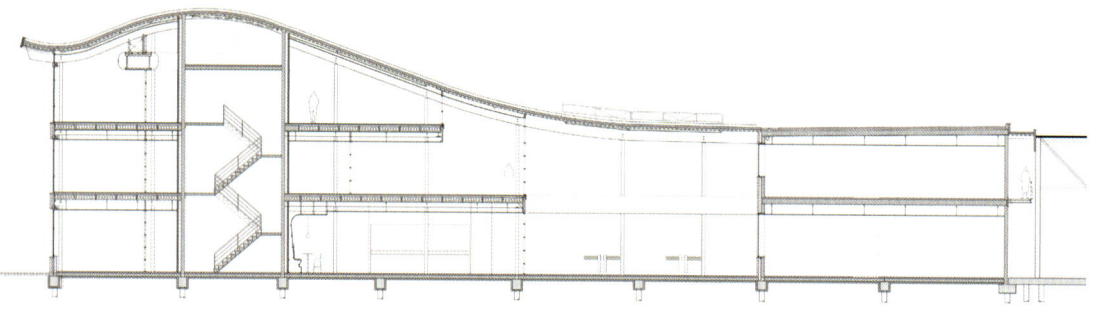

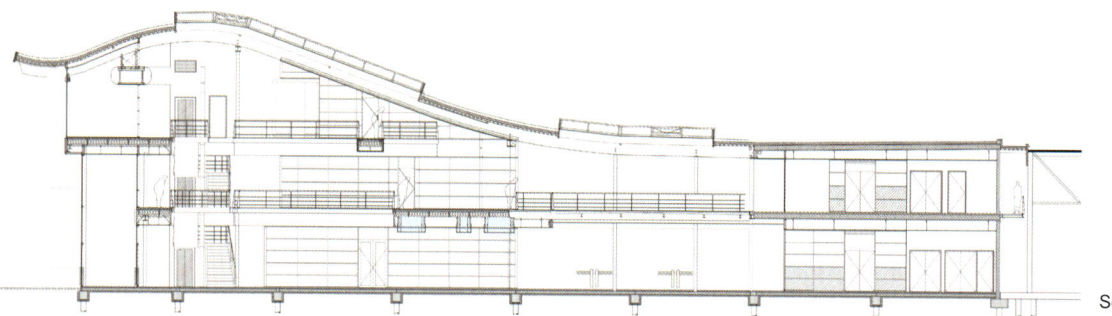

Sections

4

1. Laboratory
2. Conservatory
3. Atrium
4. Restaurant
5. Conference room
6. Offices

Ground Floor Plan

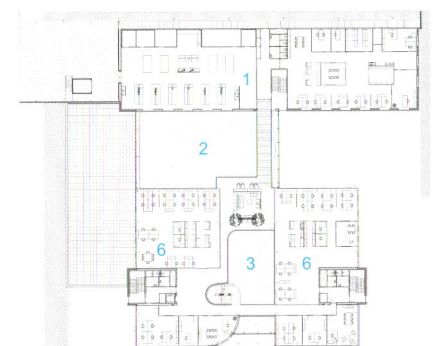

First Floor Plan

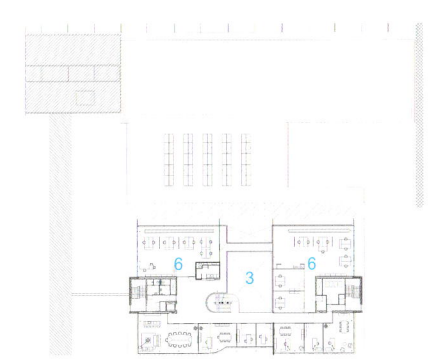

Second Floor Plan

8

1. Office and laboratory are separated by a conservatory of 450m²; a walkway ensures that both buildings are connected.
2. The extending large undulant roof
3. The large roof stretches out from the lab building across all of the facilities.
4. View of the entrance façade
5. The building is transparent with lots of light and space.
6.7. The special staircase to first floor is like a waving reeds.
8. The wooden beams in a wave shape on the ceiling
9. The space is highly transparent, light and extroverted; it has its "eyes on the world" and is inviting to employees and visitors.
10. Laboratory

TIPS:

1. Functions of a roof

Some roofing materials, particularly those of natural fibrous material, such as thatch, have excellent insulating properties. For those that do not, extra insulation is often installed under the outer layer.

Concrete tiles can be used as insulation. When installed leaving a space between the tiles and the roof surface, it can reduce heating caused by the sun.

Forms of insulation are felt or plastic sheeting, sometimes with a reflective surface, installed directly below the tiles or other material; synthetic foam batting laid above the ceiling and recycled paper products and other such materials that can be inserted or sprayed into roof cavities.

2. Solar roofs

Newer systems include solar shingles which generate electricity as well as cover the roof. There are also solar systems available that generate hot water or hot air and which can also act as a roof covering. More complex systems may carry out all of these functions: generate electricity, recover thermal energy, and also act as a roof covering.

>>> Solar systems can be integrated with roofs by:

• integration in the covering of pitched roofs, e.g. solar shingles.
• mounting on an existing roof, e.g. solar panel on a tile roof.
• integration in a flat roof membrane using heat welding, e.g. PVC.
• mounting on a flat roof with a construction and additional weight to prevent uplift from wind.

3. Gallery of significant roofs

a: Imbrex and tegula tiles on the dome of Florence Cathedral.
b: The marble dome of the Taj Mahal.
c: The polychrome tiles of the Hospices de Beaune, France.
d: The glazed ceramic tiles of the Sydney Opera House.
e: The copper roof of Speyer Cathedral, Germany.
f: The lead roof of King's College Chapel, England.
g: The glass roof of the Grand Palais, Paris.
h: The white Teflon-coated fiberglass membrane of the Centre Pompidou-Metz museum, Metz.

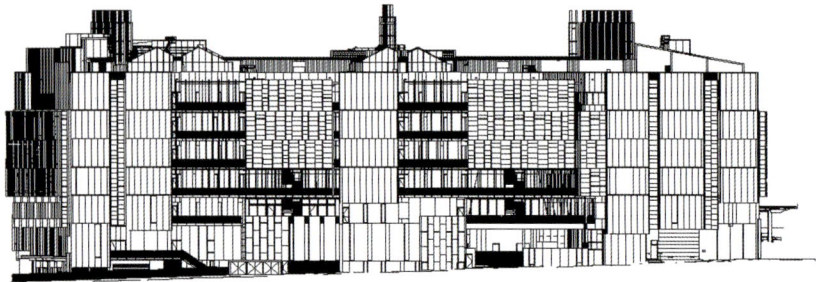

East Elevation

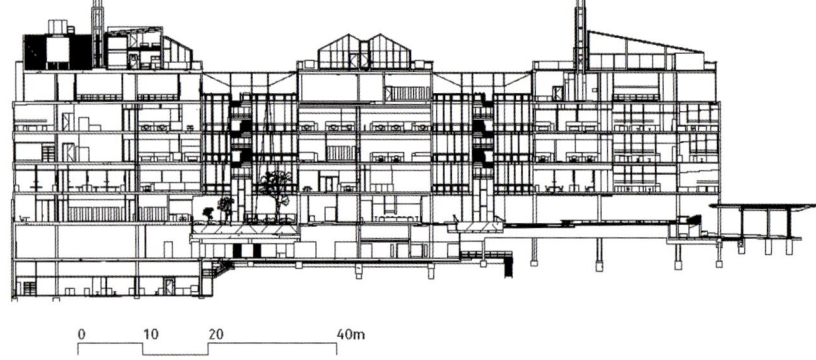

0   10   20   40m

Section

048 — *Demonstration on the Characteristics of Modern Science Buildings*

2

"Without wall" is the embodiment of the open design element

| Ecosciences Precinct

Location:
Brisbane, Australia
Architect:
HASSELL
Construction Area:
50,000 m²
Completion Date:
2010
Photographer:
Christopher Frederick Jones

▪ As Ecosciences is located in the subtropical region of Australia, one of the major challenges of the project was filtering the harsh sunlight. The solution was to envelope the building in a veil of perforated aluminium sun-screen, protecting the laboratory and courtyard spaces while establishing the external aesthetic of articulated and perforated skin.
▪ "Without walls" is a major innovation, shedding spatial and operational barriers to optimise collaboration, knowledge exchange and sharing.

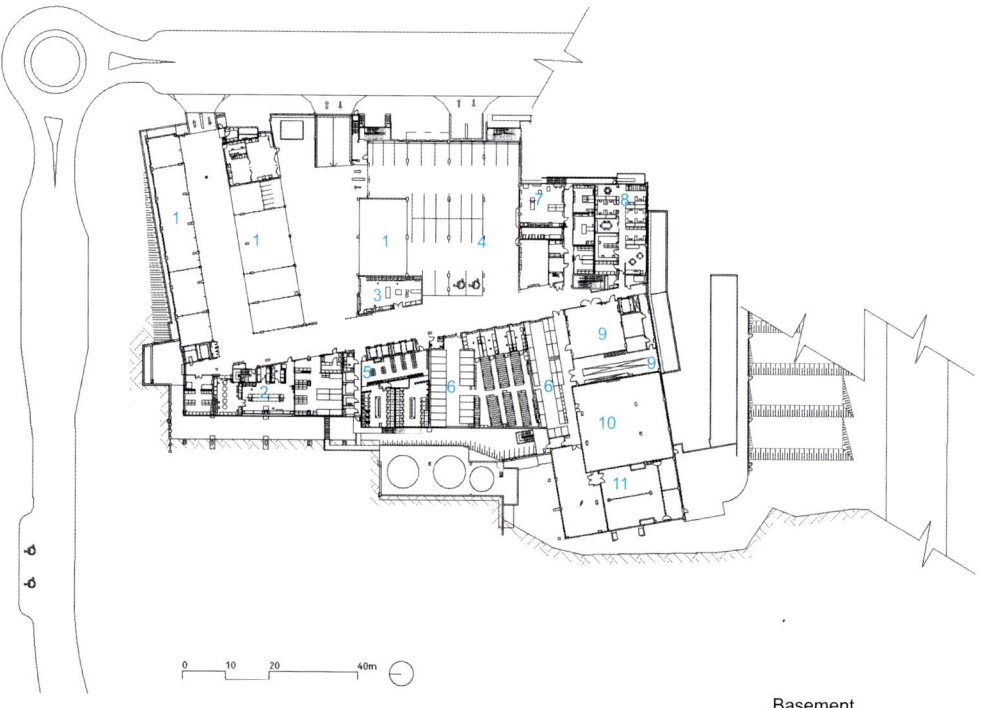

1. Parking
2. Marine sample processing
3. Workshop
4. 4WD parking
5. Lockers and shower
6. Field gear storage
7. Timber pilot plant
8. Facility management offices
9. Electrical plant
10. Chiller plant
11. Generators

0 10 20 40m

Basement

1. The building makes use of perforated aluminium veil to create filtered light in large indoor spaces.

2. The eco lab as a whole employed a number of sustainable strategies focusing on northern orientation, passive solar design, extensive sun-shading, advanced building control systems, energy, water and waste efficiency.

3. The internal courtyard creates a mini city within the facilities, which encourages scientists and researchers to get out of their labs to meet with each other.

4.5. Staff interaction area

6. Efficient layout for internal circulation

7. The Precinct is broken up into three zones – office, laboratory and support – accommodating varying group sizes and functions in a generic.

8.9. Stimulating, open and transparent research spaces form an ideal environment for creative professionals who demand not only state-of-the-art workplace facilities but also urban amenity and diversity of activity in a safe, vibrant setting.

10. The flexible and adaptable configuration is designed to accommodate change over time.

1. Entry courtyard
2. Café
3. Reception
4. Security access control
5. Foyer
6. Seminar rooms
7. Seminar courtyard
8. CSIROSEC labs
9. CSIROSEC courtyard
10. Library
11. Staff interaction area
12. Staff courtyard
13. Offices
14. Laboratories
15. Pedestrian spine

0   10   20   40m

Site & Ground Floor

1. Passenger lifts
2. Staff interaction area
3. North-south street
4. Atrium
5. Goods lift lobby
6. External circulation
7. Courtyard
8. Office
9. Laboratory
10. Laboratory support

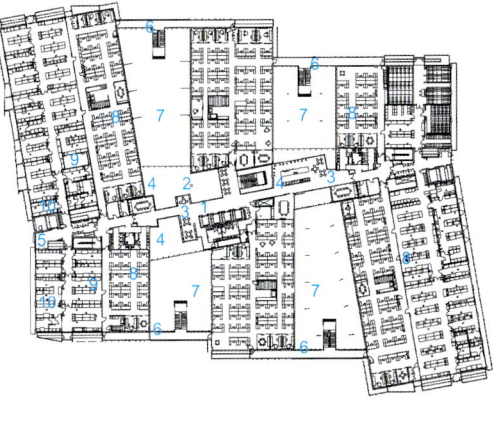

Level 1 and Level 2 Floor Plan

9

TIPS:

1. Local Climates:

Brisbane has a humid subtropical climate with warm to hot and humid summers and dry, moderately warm winters. Humid subtropical climates normally lies on the southeast side of all continents, generally between latitudes 25° and 40° north and tend to be located at coastal or near coastal locations. However, in some cases the climate extends well inland, most notably in China and the United States.

2. Perforated Material:

>>>Advantage:

• Acoustic performance
• Weight reduction
• Radiation containment
• Separation
• Heat dissipation
• Transparency
• Anti-Skid

>>>Applications:

• Interior Design
• HVAC
• Noise Control
• Security Ceilings
• Sunscreens
• Building Facades
• In-Fill Panels
• Ceilings
• Furnishings
• Stair Treads and Risers
• Screening and Fencing

10

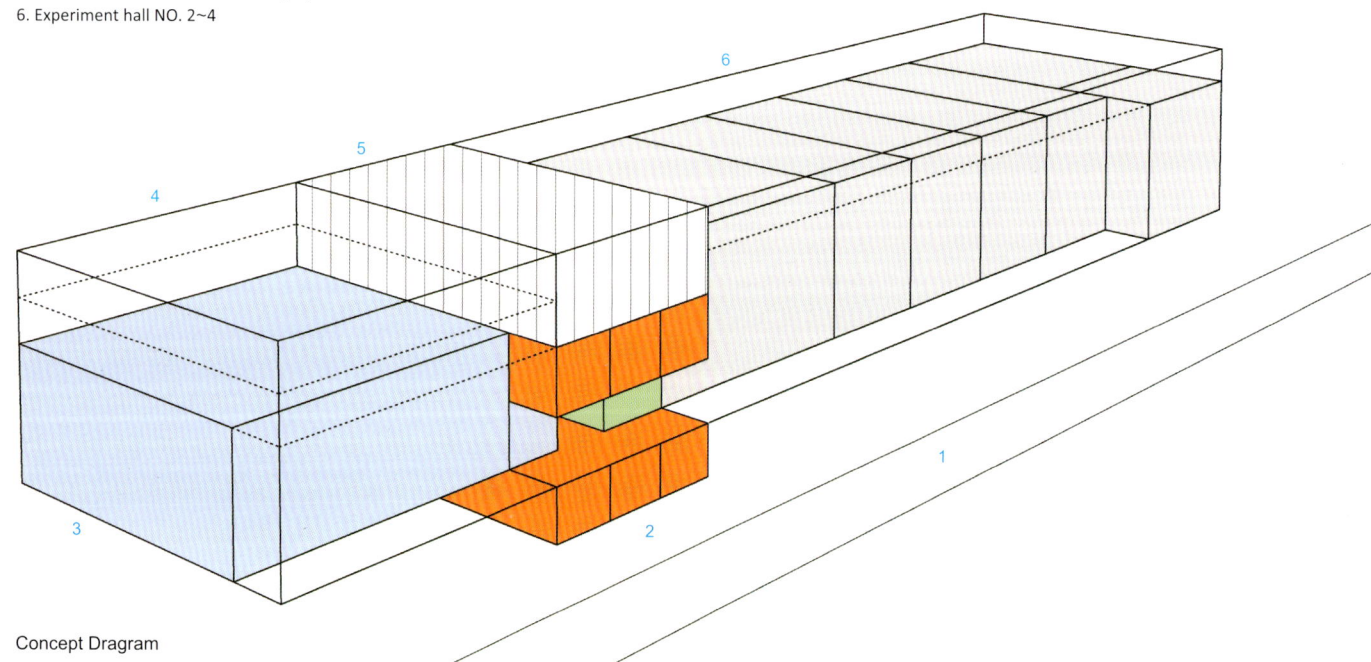

1. Street
2. Main entrance
3. Experiment hall NO.1
4. Roof equipped with heat exchange and solar heat
5. Technical levels and rooms for employers
6. Experiment hall NO. 2~4

Concept Dragram

Exterior materials create visual splendor
| e.on Energy Research Centre der RWTH Aachen

Location:

Aachen, Germany

Architect:

fischerarchitekten GmbH & Co.

Construction Area:

2,650 m²

Completion Date:

2009

Photographer:

Olaf Mahlstedt, Peter Hinschläger

- The building features a clear, well proportioned linear form which pleasantly contrasts the surrounding functional architecture.
- The designer uses various materials and colours in the exterior design to provide the audience a cool and graceful visual feast.
- In construction design, the designers carefully considered the building's functions, improve the exhaust rate, enhance the interior lighting and add energy-efficiency and environment protection advantages.

- To meet the requirements of the client, the design of hall improves the flexibility (to conduct different experiments). The glazing enables the building to integrate with the surrounding highway landscape.
- The designers uses U-profiled glazing: The dressing and ancillary rooms in the central zone are housed in an area separated by U-profiled glazing which conveys a solid character through illumination.

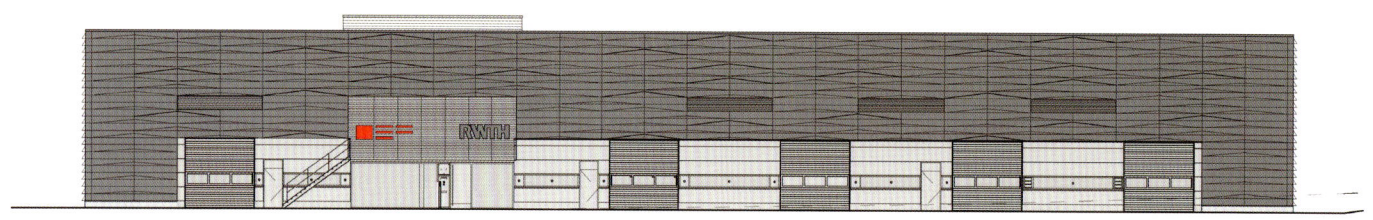

Northwest Elevation

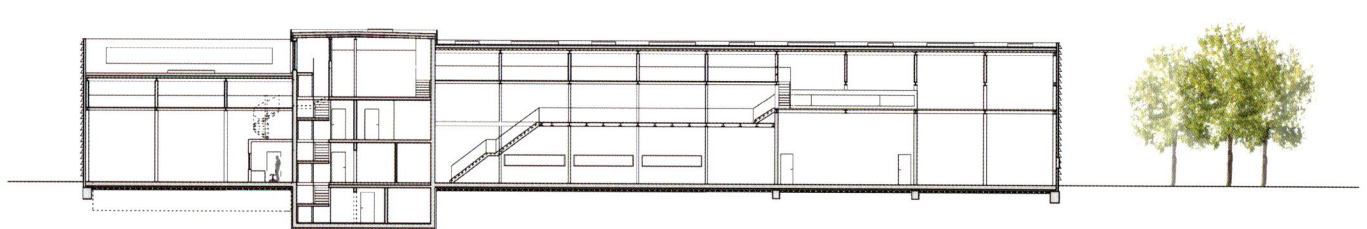

Longitudinal Section

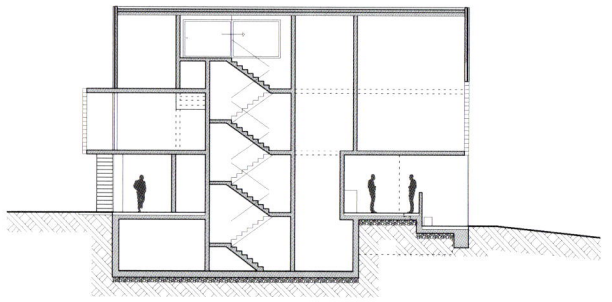

Cross Section Entrances Kitchen Stairs

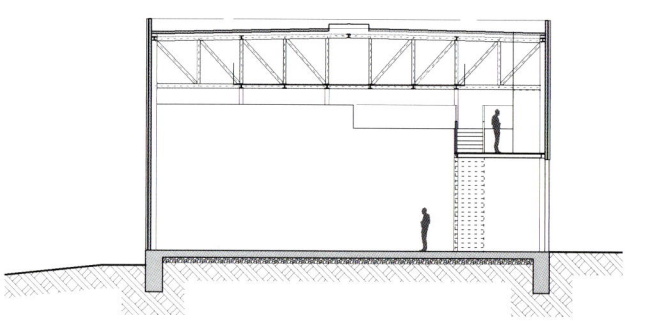

Cross Section Experiment Hall

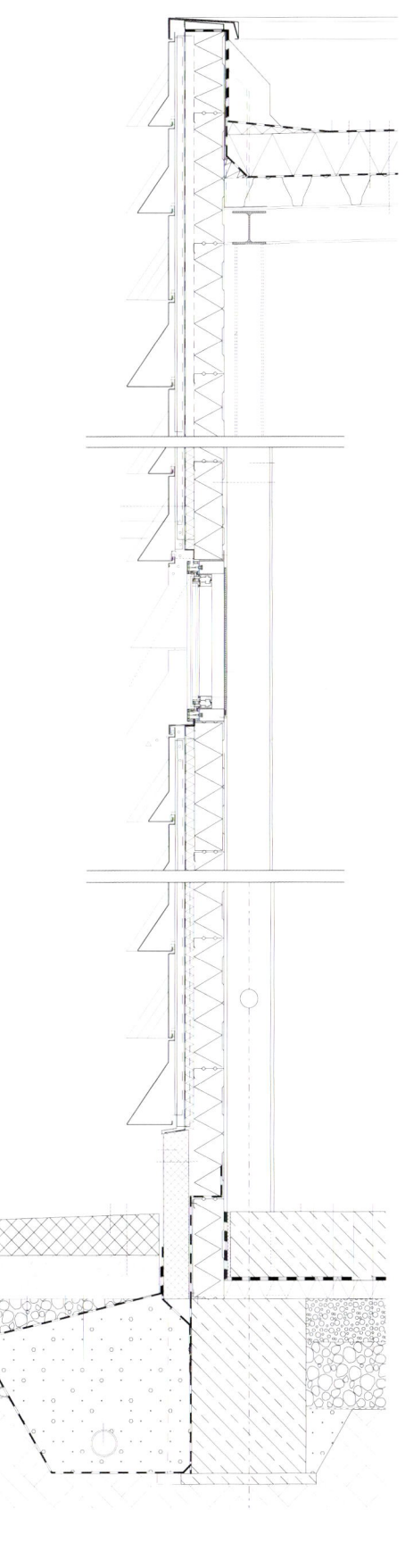

4

Façade Details

5

TIPS:

Award:
2010 Awards for excellent buildings from the Federation of German Architects, Aachen.

Award description:
To promote the quality of the built environment in cities and regions and to stimulate the public discussion about architecture and urban design, the BDA (Federation of German Architects) regularly awards architecture prizes. Every three years the local groups of the BDA in North Rhine-Westphalia commend the "Auszeichnung guter Bauten" (Awards for excellent buildings). These awards are meant to contribute to the improvement of quality of design and construction and to set quality standards for contemporary architecture.

Award Reason:
The building features a clear, well proportioned linear form which pleasantly contrasts the surrounding functional architecture. Doing this the whole structure exudes concentration and discipline in dealing with the technical and functional aspects of the brief. The attention to detail and material in the interior design of the staff areas of the building is striking. The structure is clad with multiple folded dark metal panels that lend an elegance and effortlessness to the building which one does not expect at this location.

1. The whole structure exudes concentration and discipline in dealing with the technical and functional aspects of the brief.
2. The metal cladding consists of profiled panels which create a flickering image of shadows on the façade. The black grey RAL 7021 supports the shadow effect created by the panel edges and stands for calmness and elegance.
3. The entrances and gates to the halls are Aluminium white RAL 9006. They form a strong contrast to the dark main structure and emphasise the function of these areas of the façade.
4. The external walls are a double leaf construction made of insulated metal panels, a ventilation layer and external metal cladding.
5. The attention to detail and material in the interior design of the staff areas of the building is striking.
6. A row of 2.0m × 2.5m large skylights is integrated into the central part providing smoke vents and also natural light.
7. In parts of the hall the ceiling was constructed as a lightweight metal grid between the trusses.

6

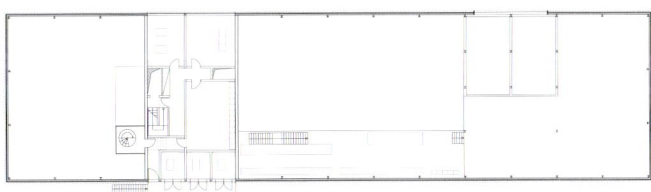

Leve 1 Plan

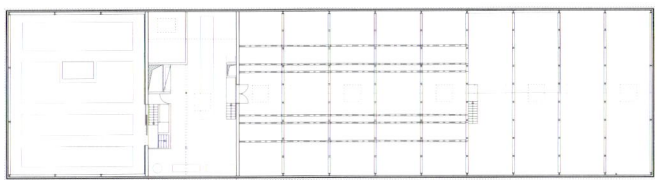

Leve 2 Plan

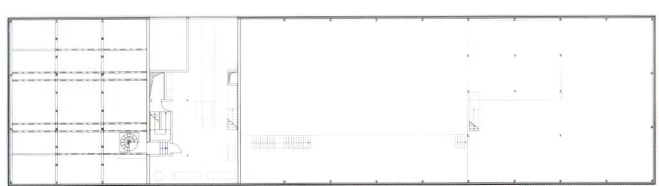

Leve 3 Plan

7

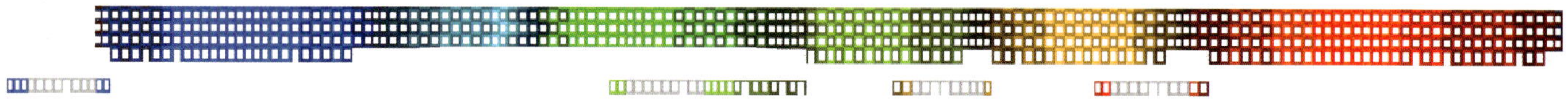

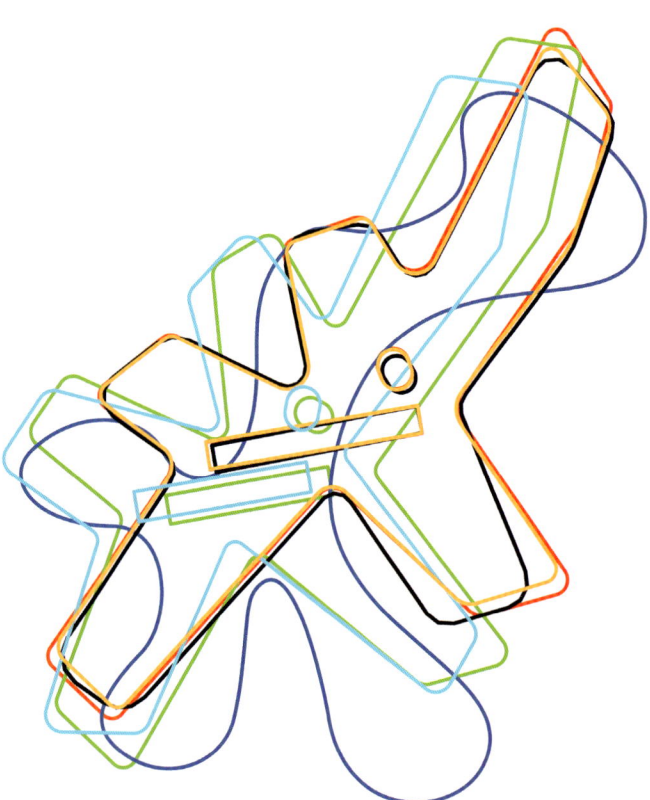

Wettbewerb Phase 1
Wettbewerb Phase 2
Wettbewerb Phase 3
Feasibility Study 2005
ES-Bau
Realisiert

Utilisation of façade structure reduces energy use
| Biomedical Research Centre Seltersberg of Justus-Liebig-University Giessen

Location:
Giessen, Germany
Architect:
Behles & Jochimsen architects BDA
Construction Area:
27,714 m²
Completion Date:
2011
Photographer:
Marcus Bredt

▪ Reducing energy use through the structural design of façade and some technical measures: The frames around the windows are made from aluminium anodised in all available shades by means of the Sandalor method.
The fenestration allows for an optimal exposure of the work area while unnecessary energy gains are being avoided.
On the outside, perforated metal sheets indicate the position of the non-transparent flaps used for ventilation purposes.
Passive and active, structural and technical measures help to significantly reduce the amount of energy needed.
The compact structure and great depth of the building result in a favorable A/V-ratio (A:

area of building envelope surface, V: volume of the building).
▪ The building forms five "feet", representing different functional zones and connected by public corridors, making the whole plan transparent.
▪ In interior design, "separate corridors" are used in the spatial division of groups of classified laboratories.
▪ Interior layout features: The building is structurally characterised by an outer layer of rooms parallel to the façade and an inner zone, which holds dark rooms and technical appliances. Laboratories and offices are based on the same module. The flexible structure of the plan is interspersed with communicative zones.

2

1. Flooring (laboratories/offices)
   PVC
   T=5mm (including gluing),
   80mm chamfer circumferential
2. In-situ concrete ceiling (grouting)
   T=22cm
   precast upstand as stop end formwork
3. Façade cladding
   mirrored, coloured glass single layer toughened glass (ESG) or multi-layer
   laminated glass (VSG)
   fixing by undercut or bonding
   also outer cladding ventilator
   R+W
4. Insulation/sealing/rear ventilation
   T= 12cm (Insulation)
   D= approx. 9cm (rear ventilation)

5. Filigreed ceilling
   T= 8cm
6. Window frame
   H= approx. 220cm, W= approx. 300cm
   Including wall duct in the window profile intergrated, aluminum blank
   Connecting BK by riser duct
7. composite window
   H= approx. 200cm, W= approx. 210cm
   rotary wing for cleaning and revising openable
   internal glare protection (roller with electrical Driver)
   external sun protection (blind with electrical Driver)
   aluminum blank
8. ventilation wing
   H= approx. 200cm, W= approx. 60cm
   Fitting model
   tilt & turn, aluminum blank

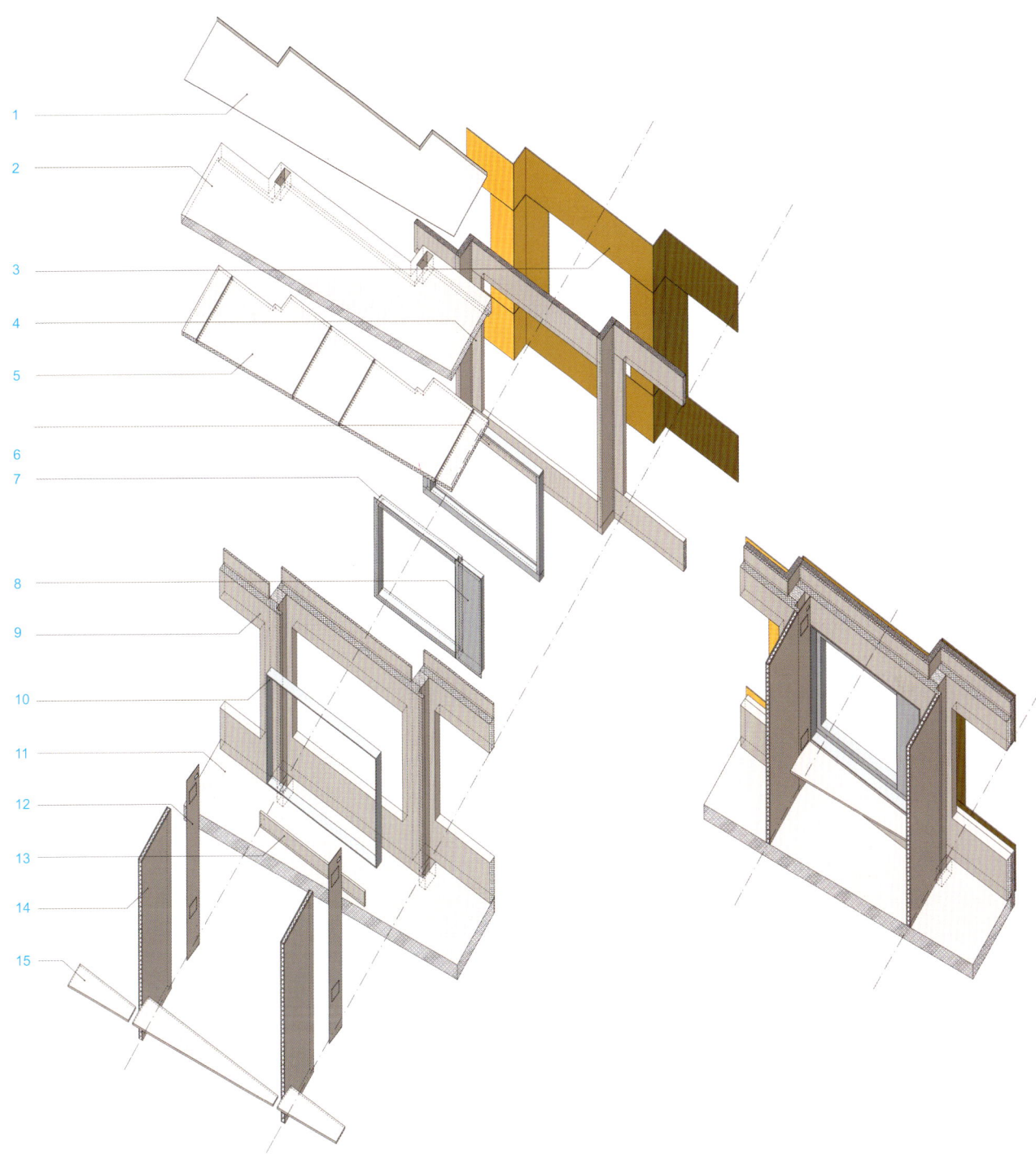

1

2

3

4

5

6

7

8

9

10

11

12

13

14

15

9. Prefabricated component
   T=17cm, grouting in in-situ concrete("in-situ concrete columns" between precast wall panels), angle variable
   Including concrete upstand as permanent formwork for filigree in-situ concrete ceiling
   Internal coating white
10. reveal frame
    aluminum blank
11. storey ceiling above first floor
    T=variable (25cm~50cm)
    precast upstand as stop end formwork
12. Manhole cover
    gypsum board
    with opening for ELT and inspection opening
13. Radiator
    Single-layer panel radiator powder-coated white

T=6.5cm, H=30cm
Length variable
14. partition wall
    gypsum baseboard, painted white,
    according to protection and lavoratory orders
15. Window worktop
    T=4cm
    including console
    wooden composite with HPL coating, white

3

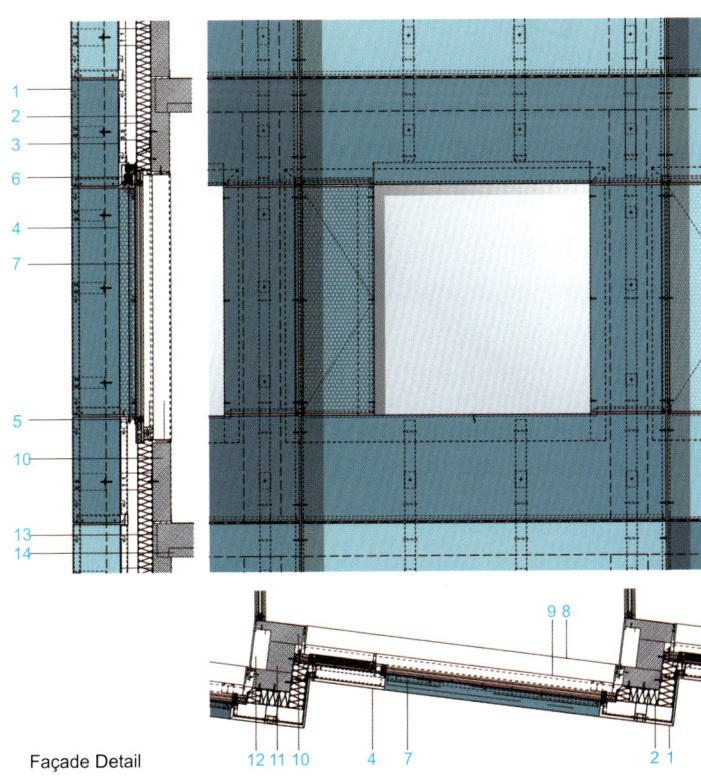

1. Ventilated curtain façades as bolted suspending façade with in the SANDALOR process coloured anodised aluminum sheets(D=3mm)
in 32 colour tones
2. Mineral insulation WLG 035
3. Substructure made of vertical mounted aluminum
Sheet coverings
U-Profiles mounted for the flanking sheet coverings
4. Coloured anodised perforated sheet (free crosssection 60%) in front of the ventilation windows
5. Covered lying drainage channel
6. Covered lying cable guided solar shading with light steering effect
7. Thermal insulated aluminium window element with Insulating glass-fixed glazing in the revision area and flanking mouted ventilation windows with panel infill (block windows); Surface of window elements outside: alu-bright rolled, inner: silver anodised (E6/EV1)
8. Adjustable aluminium frame made of extruded profiles
9. Wall duct (fasten by vertical installation shaft) in E6/EV1
10. Supporting reinforced concrete precast part shear walls, D=17cm
11. In-situ concrete grouting to the frictional connection to the shear walls
12. Installation shaft
13. Opening of the shear walls as support for filigreed ceiling tiles
14. Filigreed panelled ceilings (D=80mm) with concrete layer (smooth in-situ concrete, D=220mm)

Façade Detail

1. Finger A
2. Finger B
3. Finger C
4. Finger D
5. Finger E
6. Entrance hall
7. Finger foyer
8. Lab rooms
9. Preparation area
10. Course room

11. Seminar rooms
12. Small lecture hall
13. Large lecture hall
14. Cafeteria
15. CIP Cluster
16. Storage
17. Porter
18. Lounge
19. Workshop
20. Laboratory animal house

Site & Ground Floor Plan

1st Floor Plan

2nd Floor Plan

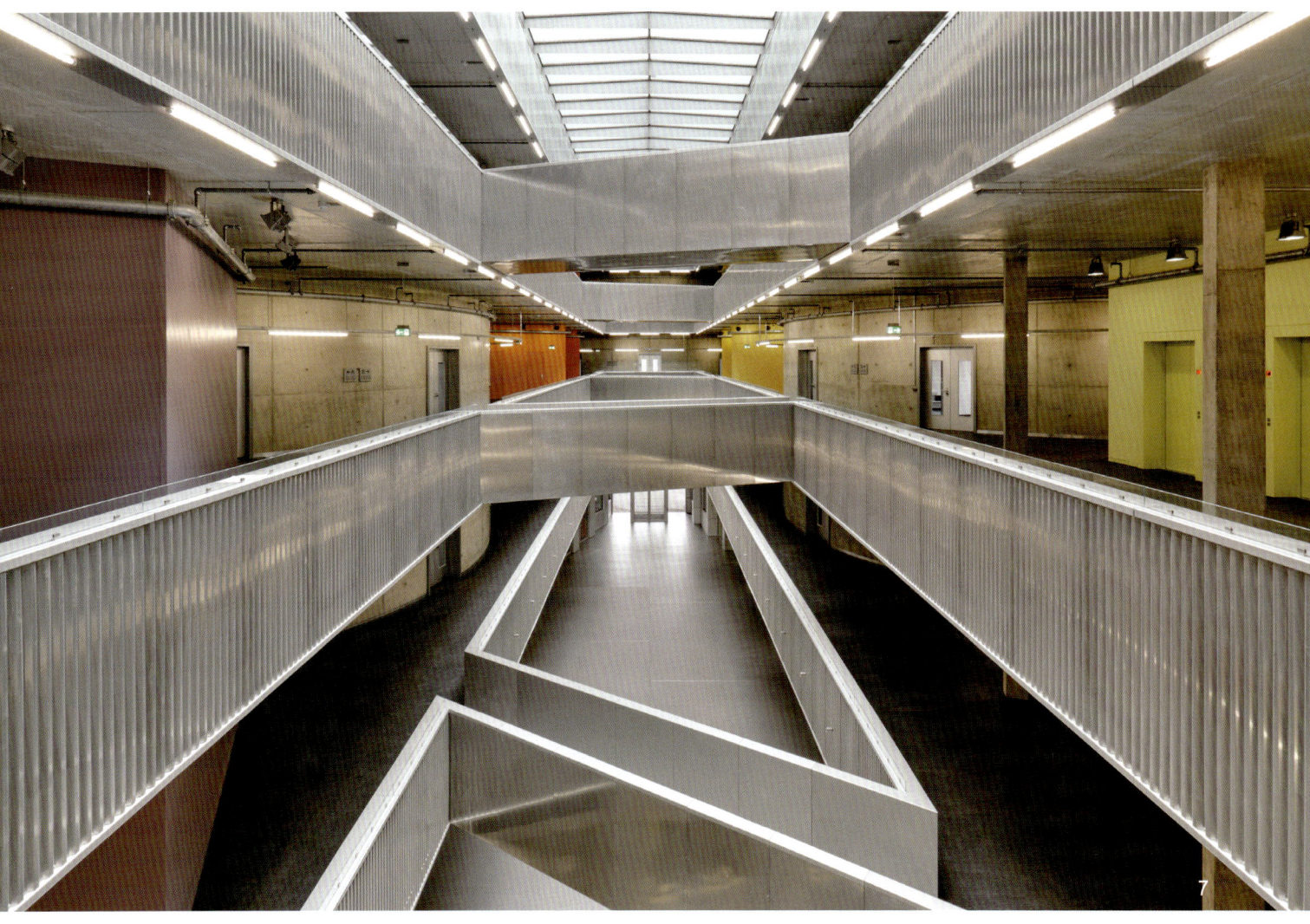

7

1. The building is completely autonomous and generates space to all sides through its five building "fingers". The design respects the building lines of the adjacent streets and lines up with the other buildings surrounding the future Campus Hill.

2. The bent and scaled façade produces a varied spatial experience in circumambulating the building. The five building fingers obtain their own color identities through coloured walls in the core zone; orientation within the building is thus facilitated.

3. The building has a rainbow-like façade made of scaled aluminium frames in 32 shades. Each building transitions between gradients of colour and brightness, while the aluminium casts various reflections as daylight shifts. The buildings are arranged like five "fingers" extending from the middle court. The façade consists of staggered elements that map the main axes with a width of 3.45 m.

4. Details of the entrance

5. On the ground floor, wall panels in the colours of the respective building fingers are introduced in the foyers in order to improve the acoustics.

6. Structural walls and ceilings in the public areas are done in exposed concrete.

7. Inside the central atrium, the ground floor hosts ample public spaces including a passageway for pedestrian traffic between the campus and city.

8. Lecture halls and seminar rooms, CIP-cluster and cafeteria as well as the training laboratories are organised in separate entities, which are attached to the main hall.

9. Laboratories and offices are based on the same module. This makes it easier to adapt the layout to future changes and complies with the users' wish of interspersing the laboratory zones with offices.

10. Transparency to the adjacent areas like the lecture hall makes the hall into a passage that allows for a variety of views into and through the building.

11. The creation of separate corridors for groups of classified laboratories is possible at almost any place without impeding the use of the other areas. The frames around the windows are made from aluminium anodised in all available shades. The fenestration allows for an optimal exposure of the work area while unnecessary energy gains are being avoided.

12. The building is structurally characterised by an outer layer of rooms parallel to the façade and an inner zone, which holds technical appliances.

8

TIPS:

1. What is anodised aluminum sheet?
Anodised aluminum sheet is a metal product consisting of aluminum sheeting exposed to an electrolytic passivation process that imparts a tough, hard-wearing protective on its surface.
http://www.wisegeek.com/what-is-anodised-aluminum.htm

2. How to paint-over anodised aluminum?
http://www.ehow.com/how_5943805_paint-over-anodised-aluminum.html
http://www.ehow.com/how_6462194_anodise-aluminum-home.html
http://w3.uwyo.edu/~metal/anodising.html

3. The advantages and disadvantages of anodising: Painted through electrolytic colouring; anodising could significantly improve the corrosion resistance, surface hardness and abrasive resistance of aluminium alloy; favourable decorative feature. However, anodising process also produce PFCs (CF4•C2F6), which will damage the ozonosphere. Many developed countries have strict restrictions on electrolysis of aluminium.

9

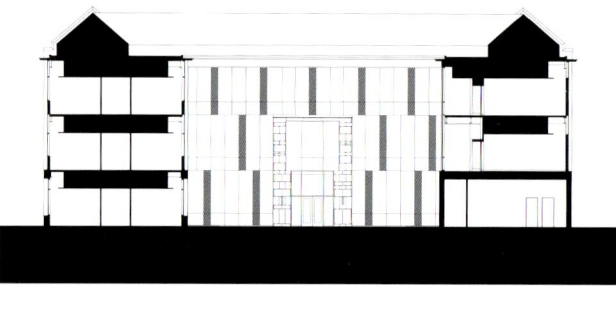

1. Flanked by the mass of the stone stair towers, the glazed interior court is an illuminated beacon at night, inviting discovery within. Transparency through the building is most evident at the fully glazed interior courtyard.

2. The entrance to the Doering Centre is centred on axis with the new student centre.

3. The stone and brick along the perimeter relate to the materials and detailing of the surrounding traditional architecture.

4. The north façade is a glass curtain wall, embraced by the east and west wings.

5. The building atrium extends through core of the building providing a visual connection and a dynamic space for interaction and communication. Cherry, a locally harvested wood, was used throughout the space and corridors to warm the interior spaces.

6. Classrooms and lab spaces were designed to be fully transparent, inviting investigation and extending the academic environment outside the learning spaces.

7. Folding marker boards were designed into all labs and classrooms providing both a teaching aide and privacy when required within the learning spaces.

8. Light filled corridors allow for student interaction and provide a visual connection to the exterior and into the classrooms and lab spaces.

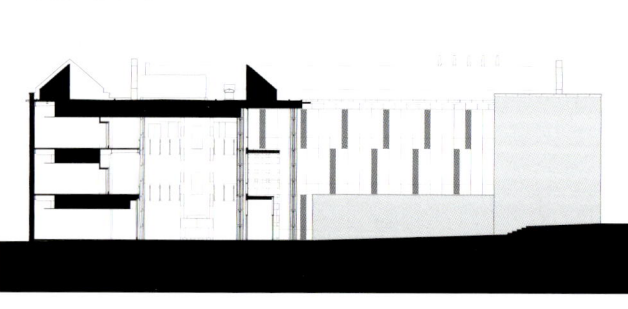

0 8 16   32

0 8 16   32

Sections

Façade stone materials and the installation methods of rain screen reduce the cost

| Bryn Athyn College Grant R. Doering Centre for Science & Research

Location:
Bryn Athyn, USA
Architect:
Spillman Farmer Architects
Construction Area:
3,623 m²
Completion Date:
2009
Photographer:
Steve Wolfe Photography

- The inspiration for the design came from the natural world in the form of the geode, a rock formation characterised by a rough stone exterior and crystalline interior.
- The site endows the project with deep historical and cultural deposits:
The institution is nestled within the small town of Bryn Athyn Pennsylvania, home to architectural masterpieces including Bryn Athyn Cathedral and historic mansions Glencairn and Cairnwood, all of which have close physical and historical ties to the college.

- The exterior and masonry features:
Exterior masonry and glass façade respect the proportions and scale of the existing campus buildings while the stone and brick details draw from the rich architectural history of the college and also the architectural vernacular of the early 20th Century Philadelphia suburbs in which the campus is located. In fact, the stone used in the building was sourced from the same regional quarry as the stone used in the Cathedral.
- The designers use many economical

techniques to facilitate maintenance and to reduce the operational cost.

▪ The slate rain screen is attached using a mortarless spring and clip system, an economical technique (as are many others in the building including the elevator mechanics and plumbing valves and piping) that demonstrates how a traditional building material such as slate can be truly timeless in application. The strategy has the added benefit of providing easy access for maintenance staff and reducing operational costs.

▪ The arrangement and design of other ancillary facilities:

Fume hood: Fume hoods are purposefully located along the interior common walls to preserve views out and maintain the transparency.

Laboratory equipments: Even the placing of the chemical prep rooms, a difficult programme element due to the functional nature of the space, is transformed into an opportunity by being strategically located between two labs.

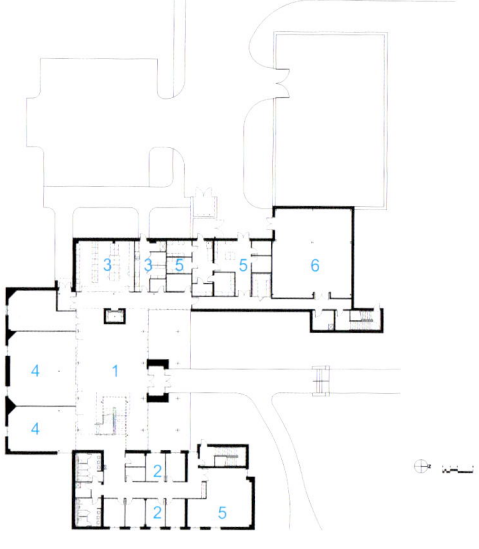

Ground Floor Plan

1. Gallery   4. Classroom
2. Office   5. Storage
3. Lab   6. Mechanical

1st Floor Plan

1. Open to below   4. Prep room
2. Office   5. Lounge
3. Lab

TIPS:

LEED
>>> How to certify a building project
1. Choose
2. Register
3. Submit
4. Review
5. Certify
>>> LEED certification involves five primary steps:

1. Determine which rating system you will use and prepare your certification application. Applications differ depending on your building type and the LEED credits you decide to pursue.

2. Register your project. The registration fee for a project is $900 for USGBC members and $1,200 for non-members.

3. Submit your certification application and pay a certification review fee. Fees differ with building type and square footage.

4. Await the application review. Review processes differ slightly for each building type.

5. Receive the certification decision, which you can either accept or appeal. An affirmative decision signifies that your building is now LEED certified.

The Requirements of Natural Environment for Research Buildings

1. The Relationship with the Surroundings

Until the seventies, research facilities – single academic institutes, research centres or industrial facilities alike – were mainly developed on detached suburban sites to prevent dangerous effects of toxic emissions and public nuisances such as noise from machinery or traffic. However, with increased regulations for emissions that contemporary buildings have to comply with, these restrictions are now superseded.

Furthermore, laboratory use of toxic substances could be reduced drastically through methods of measuring that are more precise today than ever imagined. Many dangerous substances are replaced by less toxic chemicals; in some cases, laboratory experiments are completely replaced by computer simulations. Hence, integration of industrial and scientific facilities into the urban context has become a reality.

However, recent scientific developments in the fields of nanotechnology have given rise to a whole new generation of machinery and equipment (microscopes, tomography, work benches, etcs.) that is highly sensitive to electromagnetic, seismic or acoustic influences. These influences have to be thoroughly considered, analysed and incorporated into the planning process on a case-to-case basis. They may even include unobtrusive factors such as rivers (low frequency noise of ships' bows or screws), distant tramlines (vibrations depending on the rail construction and building ground, or potential electromagnetic fields). Such factors may lead to an overall revision of the choice of location or specific on-site measures concerning foundation work or screening. To address these issues, architects may turn to historical examples, and house special equipment in separate metal-free structures (for example, made of timber). Today, the rule is: proximity is possible, provided the neighbouring buildings don't disturb!

2. The Control of Pollution

(1) **The storage of hazardous chemicals.** Laboratory facilities designate an area or areas for collecting and storing hazardous chemical, biological, and radioactive wastes before disposal. The disposal area should be located with reasonable proximity to the elevators which connect to the loading dock area for convenience of waste disposal. General waste consisting of paper and glass should be stored in a separate area of the facility not associated with hazardous waste.

(2) **The disposal of chemicals.** Pouring chemicals into a drain that flows directly into the public water system is not permitted. Chemicals must be handled locally in the lab or with dilution tanks in or near the build-

ing. Local handling is the most affordable approach: the researcher pours the chemical into a specific container that is later picked up by a waste-management staff person or by a vendor. If chemicals are allowed to be poured down the drain, then all the drains must be constructed with chemical-resistant piping, which can be very expensive. The holding tanks will take up a few hundred feet, at a minimum, at the basement level.

(3)**The purification of wastewater.** Reclaimed wastewater is an option in limited circumstances, when a laboratory has access to municipal wastewater that has been treated to a secondary disinfection level, or when treated wastewater can be generated cost effectively on site. Reclaimed wastewater might be used for some nonpotable applications, such as cooling tower make-up.

(4)**The disposal of contaminated air.** Contaminated air from the fume hood enters the unit and passes through a packed bed then through the liquid spray section, a mist eliminator and then into the exhaust system for release outside the building. Many laboratory designers and facility operators are beginning to consider some type of scrubber for their fume hoods. This is due, in part to concern for the environment, environmental health, and increasing government regulations. There are many types of pollution control devices for laboratory fume hoods. These include liquid scrubbers, adsorbers, and particulate filters.

(5)**The arrangement and insulation of drainage pipeline.** The design discharge, pipeline calculation and the choice of tubing should meet the requirements of a research building. Thus, the drain system should fit to the quality, rate of flow and regular pattern of the wastewater, with consideration of indoor and outdoor drain conditions. The wastewater with toxic and hazardous substantial should be separated from domestic sewage and other wastewater. The liquid waste with a comparable pure substantial and valuable regent should be recycled.

(6)**The control of noise.** The design should ensure that ambient laboratory noise levels emanating from installed systems and associated equipment will not preclude effective communication at normal voice levels. Generally, this means that 55 dBA throughout the area, or 60 dBA immediately adjacent to noisy equipment such as fume hoods, should not be exceeded. Simple measures such as the installation of flexible ducting to fans, sound isolation of ducts and motor mount attachments, proper location of HVAC equipment and the operation of systems within efficient load ranges should alleviate this potential concern.

Designers must define the nature of the envelope of the rooms, starting with such structural elements as floors and walls. Those elements will often be made of reinforced concrete due to both noise and vibration control purposes and fire and safety purposes. The nature of non structural envelope elements such as partitions window and doors must be defined according to the acoustical targets as well as the safety requirements.

The projected technical and mechanical layout must be investigated in order to ensure it does not act as a sound bridge between spaces, nor does it generate undue noise. Ceiling and wall cladding must be defined in accordance with acoustical and safety requirements, but also with such operational requirements as cleanliness.

3. Fitting to the Local Physical Geography

(1)The relationship between architecture and climate is that the building should fit to the climate, creating appropriate indoor climate and the building's climatic effects should cohere with the local climate. Therefore, in architectural planning, overall arrangement, foundation, structure, equipment, water supply, drainage system, construction and the choice of best plan, the designers should consider the local climate and provide specific climatic indicators according to the building's requirements. The most influential factors of climate is temperature, solar radiation, wind, rainfall, relative humidity, etc.

The temperature can directly decide various climate parameters of research buildings such as thermoengineering performance, heating and air-conditioning load, thus deciding the design of outbuilding envelope insulation, indoor ventilation and air-condition, etc.

The main influence of solar radiation is reflected in three levels. Luminous effect: the visible light in solar radiation can influence the lighting and interior illumination of the building; Heat effect: the sunlight comes into the interior through windows and warm the walls, thus heating the indoor air, influencing human bodies and the indoor experiment facilities; UV effect: UV can accelerate the ageing of construction materials, especially organic materials such as plastics, and cause their damages.

The influence of wind reflected in that wind load is one of the main load in architectural design, which will influence the building's economy, safety and application. The wind direction and speed are related with the layout and natural ventilation of the building. The wind will also drive the rain to wash out the building's envelope, thus causing its erosion.

The amount of precipitation and the precipitation intensity concern to the drain system design of roof, ground and underground. The rain water may permeate into the interior through the cracks on the walls and make the walls get damp inside, reducing the walls' thermoengineering performance. It also may cause the roof to deform, crack and leak; the wall may become spotted and influence the visual effect; the paint may even peel and get damaged.

The relative humidity may cause many materials to get damp and reduce their insulation performance. An over-high humidity will reduce the material's mechanical strength obviously and cause its deformation. The damp materials are easily affected by mould, which will pollute the equipments and cause their ageing once it spreads into the air and the experiment equipments.

(2)For the research buildings located in the countries and areas with frequent earthquakes, they should choose seismic resistance sites and avoid the disadvantageous sites. The open site with thick and even medium-hard soil is preferable, while the site with soft soil, liquefying soil or uneven soil is unfavourable. The building's form should be simple and regular. The centre of mass and the centre of stiffness should be as close as possible, in order to avoid forming a vulnerable part in the torsion and stress concentration during an earthquake. The design should ensure the structure's integrity and the ductility of the structural and linking parts. What's more, it should choose construction materials with good seismic resistant performance.

For research buildings, the installation methods are important for damage control during an earthquake.
• Any equipment shall be permanently braced or anchored to the wall and/or floor. This includes, but is not limited to, appliances and shelving (to be installed by the contractor) which is forty-two inches or higher and has the potential for blocking corridors or doors, or falling over during an earthquake. All equipment requiring anchoring, whether installed by a contractor shall be anchored, supported and braced to the building structure.
• All shelves shall have passive restraining systems. Shelf lips must be at least one and one-half inch high. For shelves that only store books, a rubber type sheet that you put under the books, designed specifically for this purpose, can be used in lieu of lips. The shelves themselves shall be firmly fixed so they cannot vibrate out of place and allow the shelf contents to fall.
• All compressed-gas cylinders in service or in storage shall be secured to substantial racks or, even more appropriate, sufficiently sturdy storage brackets. They shall be secured with two chains, straps or equivalent, at one-third and two-thirds the height of the cylinders to prevent their being dislodged during a violent earthquake.
• For safety and ease of maintenance, it usually makes sense to locate a safety shower, fire extinguisher, and shutoff valves at the entry alcove of each lab. Interior glazing permits easy surveillance of the laboratory. Warning signs with the appropriate symbols should be posted at laboratory entrances. There should be two means of egress from each main lab (measuring 900 sq. ft. or more). Doors should swing out of main labs for safe egress in case of emergency.
• All mechanical systems should be electronically monitored, and all safety equipment should be tested on a regular basis.
• Floor penetrations should be avoided, if possible, to prevent chemicals released during a spill or flood from traveling to the floor below. Wet vacuuming should be used instead of floor drains to contain chemical spills.
• Designers should consider placing an emergency centre in a central location on each floor, to provide easy access for everyone. An emergency

centre consolidates reagent neutralisers, hand-held sprays, first aid, and fire control equipment in one common area. The centre should contain a fire extinguisher with hanger, two 1 gal. plastic bottles, a first aid kit, a fire blanket, and a galvanised sand pail.

The Planning of Laboratory Spaces

1. The Issues that Should Be Concerned in Laboratory Planning

Programming should begin with a clear definition of the activities to be performed and the people who will use the space. Accounting for functional and psychological needs is a primary purpose of the planning process that defines an owner's functional and physical requirements for each spatial element in a building or facility.

(1)Understand the Mission and Purpose of a Facility
• Determine facility use, occupancy, and activities to be housed.
• Consider the functional needs in the context of all the other design objectives to ensure a balanced and integrated design.
• Balance the owner's and users' needs and goals for space, quality, budget, and time.
• Set owner's design objectives in the early planning stage.
• Make reference to building type guidelines.
• Look beyond the facility to understand the role the site plays in meeting the functional needs in support of the mission and purpose of a facility.

(2)Define Spatial Requirements for Occupant Activities and Equipment
• Consult all pertinent stakeholders for their requirements.
• Consult planning guides and specialists on programmed activities.
• Document all regulatory requirements, such as building codes, accessibility laws, anti-terrorism/force protection (ATFP), etc.
• Explore the possible necessity of making spaces flexible to accommodate changes in business practices, work activities, and technologies.
• Consider building operations and maintenance activities in the design of spaces.
• Look beyond the facility to understand the role the landscape plays in defining spatial requirements for occupant activities and equipment.

(3)Understand the Functional Relationship between Spaces
• Engage user groups in facilitated discussions to brainstorm solutions.
• Examine patterns of activity in facility programmes and consider how those patterns create spatial relationships.
• Account for physical security requirements in the layout of space planning.
• Consider impacts of building systems and engineering needs on spatial relationships in occupied and unoccupied indoor and outdoor spaces.
• Leverage opportunities for quality environmental aesthetics such as natural light, spatial volume, views, connection to the landscape and na-

ture, texture, and materials.
- Look beyond the facility to understand the role the site plays in the functional relationship between spaces.

(4)Pre-consider the Installation, Operation, Spatial Modifications and Changes of the Facility

- Incorporate structural and mechanical systems as integral parts of early design concepts.
- Account for structural loads (dead and live) of building systems and equipment.
- Ensure that mechanical system equipment and furniture, fixtures, and building equipment (FF&E) can actually be installed, operated, and replaced.
- Consult facility O&M personnel in the programming and early design stage.
- Plan infrastructure for flexible spatial modifications or "churn" and repurposing of the building.

(5)Enough Spaces to Accommodate System Facilities such as Information and Communication

- Determine the owner's goals and needs for spatial and mechanical support of the organisation's IT programme.
- Incorporate IT system needs as an integral part of the design concept.
- Design for configuration flexibility within workspaces that promotes occupant productivity.
- Accommodate network support and servicing requirements in the design of spaces.

(6)Other Related Issues

- Computer-based space programming applications
- Appropriate accommodation for the changing nature of work (flexibility and productivity)
- Virtual workplaces and increased use of "Hoteling" for work space
- Building Information Modeling (BIM) (defining object functionality for facility life cycle)
- Adaptability for possible change of building needs and function over time
- Though today's emphasis is put on sustainability and green building, maintain a balanced perspective with the other key design objectives.

2. Benefits of Lab Planning Module

The laboratory module is the key unit in any lab facility. When designed correctly, a lab module will fully coordinate all the architectural and engineering systems. A well-designed modular plan will provide the following benefits:

(1)It will response to the structural changes of the research spaces effectively.

The lab module, should "encourage change" within the building. Research is changing all the time, and buildings must allow for reasonable change. Most academic institutions annually change the layout of 5% to 10% of their labs. The changing nature of work means greater mobility for workers, a multiplicity of workspaces within and external to buildings, greater use of geographically dispersed groups, increased dependence on social networks – and greater pressure to provide for all of these needs and behaviors in a leaner and more agile way. Workplaces have responded with many new options, including more teaming and informal interaction spaces, more supports for virtual individual and group work, more attention to integrating learning into everyday work experience, greater flexibility in work locations, and more focus on fitting the workplace to the work rather than vice versa. Many workplaces are also incorporating spaces that encourage relaxed engagement with colleagues to reduce stress and promote a sense of community.

Support Mobility
• Consider wireless technology and mobile phones to enable workers to move effortlessly among spaces as their needs change.
• Provide a multiplicity of spaces for individual and group work.
• Provide connections to internal networks and to the Internet throughout the workplace.

Enable Informal Social Interaction
• Provide multiple places to meet and greet.
• Consider providing informal workspaces in cafeterias.
• When designing cafés and coffee nooks, locate them centrally along well traveled pathways to encourage use and interaction.
• Design the circulation system with informal communication opportunities in mind.

Design for a Variety of Meeting Sizes and Types
• Provide enclosed rooms to support groups of different sizes.
• If open informal spaces are used, make sure that they are separated from individual quiet spaces.
• Consider sharing meeting spaces among private offices.
• Provide visual display technologies and writing surfaces for group work.
• Consider the use of dedicated project rooms for some types of group work.

Support Individual Concentration
• If open spaces such as pods or bull pens are used, provide attractive acoustically sound rooms for individual concentration as needed.
• Locate concentration booths close to work spaces.
• Zone space for range of quiet and interactive needs.

(2)It will response to the size changes of the research spaces effectively.

The use of lab planning modules allows the building to adapt easily to needed expansions or contractions without sacrificing facility functionality.

A common laboratory module has a width of approximately 10 ft. 6 in. but will vary in depth from 20 ft~30 ft. The depth is based on the size necessary for the lab and the cost-effectiveness of the structural system. The 10 ft. 6 in. dimension is based on two rows of casework and equipment (each row 2 ft. 6 in. deep) on each wall, a 5 ft. aisle, and 6 in. for the wall thickness that separates one lab from another. The 5 ft. aisle width should be considered a minimum because of the requirements of the Americans with Disabilities Act (ADA).

Two-Directional Lab Module – Another level of flexibility can be achieved by designing a lab module that works in both directions. This allows the casework to be organised in either direction. This concept is more flexible than the basic lab module concept but may require more space. The use of a two-directional grid is beneficial to accommodate different lengths of run for casework. The casework may have to be moved to create a different type or size of workstation.

Three-Dimensional Lab Module – The three-dimensional lab module planning concept combines the basic lab module or a two-directional lab module with any lab corridor arrangement for each floor of a building. This means that a three-dimensional lab module can have a single-corridor arrangement on one floor, a two-corridor layout on another, and so on.

3. The Requirements for the Spatial Flexibility

Mobile casework can comprise mobile tables and mobile base cabinets. It allows researchers to configure and fit out the lab based on their needs as opposed to adjusting to pre-determined fixed casework.

(1)The flexibility of laboratory spaces
The designers should consider the installation and fixation of future equipments and the movement of various cabinets in the early stage of the design. (See Figure 1)

(2)The flexible zoning can divide the lab into working areas with different sizes. (See Figure 2)

1

Figure 1. Movable cabinets

(3)Overhead service carriers

These service carriers are hung from the ceiling. They can have utilities like piping, electric, data, light fixtures, and snorkel exhausts. They afford maximum flexibility as services are lifted off the floor, allowing free floor space to be configured as needed. (See Figure 3)

(4)The flexibility of engineering system

Labs should have easy connections/disconnections at walls and ceilings to allow for fast and affordable hook up of equipment; The engineering systems should be designed such that fume hoods can be added or removed; Space should be allowed in the utility corridors, ceilings, and vertical chases for future HVAC, plumbing, and electric needs. (See Figure 4)

Figure 2. The flexible zoning is designed for doing different kinds of work at the same time.
Figure 3. Some facilities are lifted off the floor to create more free floor space.
Figure 4. The installation of engineering system is exposed in the utility corridor to allow for fast and affordable hook up of equipment.

4. Efficient Lighting

(1)Efficient electric lighting in laboratories

The appropriate design of lighting systems is especially important in laboratories, given the intensity and significance of work carried out in laboratories and the long work hours spent by researchers. In addition, the lighting energy intensity in laboratories is up to twice that of a typical office space. Lighting energy use typically accounts for between 8% and 25% of total electricity use, depending on the percentage of lab area.

Daylight Integration – Electric lighting should always be designed as a supplement to day lighting.
Fixture Configuration – Use direct-indirect ambient lighting parallel to bench top.

There are two primary aspects of the ambient lighting fixture configuration in a laboratory:
• Beam direction: direct, indirect, or direct/indirect.
• Fixture location relative to bench top: parallel, per pen-dicular, or other.
Consider alternative ambient lighting options for movable benches.
Use task lighting.
Lamps and Ballasts – Use energy-efficient lamps and ballasts.
Controls – Use daylight controls for ambient lighting in perimeter zones.
Ensure that lighting zones are small enough to provide local control.
Use bi-level switching.
Use occupancy sensors for ambient and task lighting.
Use sweep-off lighting schedule with manual overrides.
Commission lighting controls.

(2)Day lighting in laboratories

Day lighting is the controlled entry of natural light into a building. The use of day lighting allows occupants to dim or turn off a building's electric lights to save energy. Day lighting is provided through windows, clerestories, roof monitors, skylights, sawtooth roofs, or special light-pipe systems. To save the most energy, the designer must integrate day lighting into the building's overall design, interior spaces, electric lighting system, and mechanical systems.

Increasing the amount of daylight in a space entails more than simply adding windows. Rather, designers must locate and size windows and other elements to ensure a relatively even amount of brightness in the building's interior, avoid excess heat and glare, and minimise the amount of bright sunlight that falls directly on work areas.

Day lighting within a space comes from three sources:
Exterior light reflected into a building from the ground, pavement, adjacent buildings, and other objects; Direct light from the sun and sky, which

is typically blocked from occupied space because of heat gain, glare, and ultraviolet (UV) degradation issues; and internal light reflected off walls, ceilings, and other interior surfaces. The most common day lighting approaches make use of side lighting, top lighting, and atria; other techniques can be used as well.

- **Side lighting.** In side lighting, light enters a space from windows below ceiling height, for views and day light. It is a very common day lighting technique. Horizontal strip windows are often used because they provide more uniform daylight than individual windows. Also, windows located higher in a space allow daylight to penetrate the interior of the space to a greater depth. If possible, separate windows should be used for views and for day lighting, because the optimal properties of the glazings are different for each use.

- **Top lighting.** For top lighting, daylight enters a space through vertical windows located above the ceiling line. Windows can be configured other than vertically if overheating can be avoided, such as by using specially designed horizontal skylights in deep window wells. Top lighting can be effective when windows are incompatible with the function of the perimeter walls, when interior spaces cannot easily accommodate side lighting, when the design or lighting criteria make side lighting inappropriate, or when there are security concerns.

Daylight apertures can face north or south. Baffles under the roof monitors or deep window wells can be used to diffuse and reflect light in the space. Top lighting can also be provided with stepped clerestory windows, sawtooth or roof monitors, or horizontal window wells. Light is diffused through a perforated metal ceiling; this prevents glare caused by direct sunlight on computer screens.

- **Atria.** Adding an atrium is a good way to increase the amount of space that receives natural light in a building. An atrium is often a central area one or more storeys high (depending on the building's height) with side lighting or top lighting.

- **Other techniques.** Not all buildings or sites are optimal for day lighting. A square building, or one with a long axis running north and south, is not optimal. In both cases, there is more east- and west-facing glass than is best for a daylit building. Even so, some techniques can be helpful. For example, designers can orient top lighting to face north and south. Another technique is to use vertical or egg-crate-shaped shading devices and selective surface glazing to reduce heat gain on the east and west sides while retaining some of the view to the outside.

- **Design Considerations.** Design considerations include the building's footprint and mass, shading issues, placement of walls and windows, colours of interior spaces, and glazing properties. Integrating day lighting with electric lighting, the quantity and quality of light, and codes and standards are also important considerations.

Building footprint and mass. As much as possible, specify a long, narrow footprint along an east-west axis; it is easier to daylight the north and south sides of a building than the east and west sides. Low sun angles on the east and west make shading difficult, so glazing should be minimal, especially on the west side.

Window shading. Because a well-designed day lighting system captures indirect light from the sun or sky, be sure to shade windows on the south, east, and west façades from direct sunlight. Shading options include "self-shading" windows in deep exterior wall sections, horizontal overhangs, louvres, vertical fins, and light shelves that can be integrated into the building's structure. Horizontal shading devices work well on the south façade. Architecturally, this means that the north and south façades will look different. Vertical baffles, fins, or wing walls are recommended for east and west façades if windows are needed there.

Interior colours, ceiling height, and window height. Specify light-coloured interior spaces, tall ceilings, and high windows to distribute natural light most effectively. If private offices must be along exterior walls with windows, specify a horizontal band of glass that is above eye level and adjacent to the ceiling on the walls. Also, provide a strip of glazing above shades, so occupants always have unobstructed windows even when they close their shades.

Glazing properties. Choose glazing that minimises heating and cooling loads and maximises visual comfort.

5. Safety Requirements

(1)Architectural Structure
• The architectural structure in the lab, such as walls, floors and ceilings, should select fire-proof or nonflammable materials.

• In consideration of safety, lab facilities should be single-storey or low-storey buildings. If designers have to build a multi-storey building, fire prevention and other safety issues should be carefully considered and equipped with relevant methods to ensure that when one lab is in emergency, the other labs around will be not or less affected, achieving safe evacuation and quick and effective control of the accident.

• To achieve safe evacuation, the lab building should accommodate at least two exit passageways and they should be located as far as possible. The exit passageway or access should meet safety requirements. Each lab should have two exits facing the corridor and be located as far as possible. The width of the door should be more than 90 cm. Any working point is no more than 25 m from the door. If the laboratory is not large (i.e. 6m²×6m²), one exit can be linked with the adjoining lab, but the other exit has to face the corridor. For the laboratory with a larger dimension, two or more than two exits should face the corridor or be close to the stairwell. The door facing the corridor should be double-leave door. Working places operating large amount of inflammable, explosive or toxic substances should also meet the above requirements. Therefore, once accidents occur, it is easy for evacuation or adopting relevant emergency measures.

• The corridor should be wide enough, so that people could still walk easily in the corridor when the doors on two sides are open. Except for fire extinguishers, fire hoses, gas defense tools, nothing is allowed to be located in the corridor, avoiding to influence walking and the efficiency of accident management. Rescue equipments should have their obvious signs. Designers can open a small cabinet on the wall of corridor to store rescue equipments. All the staff should know well the location of the rescue equipments. If the lab building is multi-storey, a lift is suggested to be installed to carry chemical reagent, flammable solvent, toxic substance and gas cylinder, etc. The access to the lift should also meet safety requirements, especially fire prevention requirements.

• The staircase and floor slabs in the lab building should select surface texture with some non-skid property or adopt relevant measures (i.e. non-slip mat) to meet the non-slip requirements. All the building materials, ducting and floor slabs should be chemical stable. Ceramics, glass, PVC are suitable for ducting materials, for they can meet typical lab requirements. If necessary, the floor slabs of the corridor, passageway and lab could be covered with plastic or rubber panels to avoid chemical corrosion. If permitted, the floor of labs with higher requirements could be covered with ceramic tiles. Rooms with precision instruments which require humid control can use wood floor. All the materials used in the lab building should meet certain fire-resistant requirements. The walls, doors and windows between the lab and corridor should resist at least 45 minutes of fire accident (in specified fire-resistant test condition).

• Anti-vibration: In the design of laboratory, anti-vibration is a concern and the lab should keep a certain distance from the vibration source. If the distance cannot meet the requirement, designers should adopt necessary anti-vibration methods to the equipments with anti-vibration requirement or to the whole lab building. For instance, they could use glass wool offcut to isolate vibration around the buildings or adopt vibration isolation foundation and anti-vibration benches. Typically, supporting vibration isolation equipments are used for higher interfering frequency while suspending vibration isolation equipments for lower interfering frequency and equipments requiring for higher horizontal anti-vibration levels.

• The floor slabs of the lab building should meet certain load require-ment. Typical floor load of laboratory is 200 kg/m², while the lab with larger load depends on its physical condition. The windows should be set on the north and south sides, avoiding on the west side. To avoid direct sunlight, designers could use horizontal or vertical sunshades, or curtains.

(2)Ventilation

• In the labs which require consistent temperature, the air should main-tain in positive pressure. In order to eliminate indoor waste heat and con-trol proper positive pressure, designers could install an adjustable mov-able perforated plate to exhaust indoor air.

• In the labs which work on radioactive substance, toxic substance, car-cinogen, infectious microorganism, mordant and volatile liquid, etc., in order to avoid the outward diffusion of the hazardous substances, indoor air should maintain in positive pressure and use proper devices (i.e. fume hood, exhaust equipment, etc.) to exhaust the harmful air to the outside. Before exhausting, filter material or other proper equipments should eliminate hazardous substance from the air to avoid environment pollu-tion.

• In places without chemical operations, such as corridors, meeting rooms and offices, the air should maintain fresh. Compared to chemical or micro-organism labs, these places should maintain in certain positive pressure.

• Compared to the corridor, the lab should maintain in proper negative pressure to let enough air come into the interior. The fume created in chemical reaction should be treated or diluted before exhaust. Design-ers should avoid the polluted air flow into the interior from fume hoods or cause cross contamination. All the chemical labs should install fume hoods to eliminate the hazardous fume or inflammable air (or steam), avoiding interior air contamination. In addition, fume hood provides a place for safe operation.

(3)Laboratory Benches

• The arrangement of laboratory benches should meet the requirement of emergent safety evacuation and avoid any "dead ends". Once an ac-cident occurs, wherever they are working, researchers have two different passages to choose. The paths between benches all lead to the corridor, so that staff could leave the lab as soon as possible in the accident.

• The benches can be divided into three types: island bench, peninsula bench and single-sided wall bench. Since staff could walk freely around the island benches and have a larger traffic area, island benches are convenient for safety evacuation. The proper safety clearance between benches should be more than 130 cm. But if there are special equipments (e.g. gas cylinders), the net distance should be 160cm~180cm. Because operators working on adjacent benches are "back to back", if necessary, designers could add an isolation screen made of nonflammable materials in between, to avoid chemicals to hurt the staff on the back. In this condition, the distance between benches should be increased.

• The safety clearance between benches and fume hood is typically more than 125cm. The single-sided benches leaning against the wall should also meet the requirements of safety clearance.

• In labs with natural lighting, the benches should not parallel the wall with lighting windows. Otherwise, when the operator stands with back to the windows, there will be his own shadow on the bench; when he faces the windows, there is glare.

• The height and width of the bench should be appropriate for safety operation. Depending on various conditions of different countries, the height of the bench is typically 80cm~90cm and the net width of the bench is 65cm. If there are large numbers of equipments on the bench, it could be widened. The width of the reagent bottle shelf can be 20cm~30 cm. Therefore, the width of two-sided bench can be 150cm~160 cm, while the single-sided bench 75cm~85cm.

• The benches use steel, wood, reinforced concrete as structural materials. The surface should be smooth, seamless, impermeable, wear-resisting, heat-resisting and corrosion-resisting, with enough hardness and difficult to be scratched, but also difficult to break glass containers.

(4)Lighting

① **Electric Lighting:**
• The lab should select proper lighting fixtures to ensure fire prevention.
• In a place where inflammable and explosive objects are operated, if there are only switch lighting fixtures, they should be treated as inset niche lighting and the access door should open outward, equipped with good ventilation. The interior daylighting side should be equipped with double glazing and tight sealing. As least one layer of the glazing is high strength glass. The niche lighting should be located at least 3 metres far from the doors and windows and 5 metres from the exhaust outlet.
• Labs, balance rooms, computer rooms, classrooms and offices are suitable for fluorescent lights to reduce glare. Besides, the temperature of the tube is low, the luminous efficiency is high and its service life is long. The ballast of fluorescent light should be separated from the inflammable ceiling or wall through heat-insulated, nonflammable materials. Designers should pay careful attention to the ventilation and heat dissipation. The room which requires consistent temperature should install the ballast outside.

• Lighting fixtures should have shields to avoid corrosion from chemical fume. The ground clearance of the bulb should be at least 2 metres and nothing inflammable is allowed to be stacked under it.

② Natural Lighting

• Daylight factor is the ratio of the net lighting area of the windows to the indoor floor area of the lab. In labs with high requirements in precision, the daylight factor should be 0.15; in labs with lower requirements in precision and normal offices, it could be 0.1. If there are shading buildings around, the daylight factor could be increased.

• Depth factor is the ratio of the distance between daylight wall to its opposite wall (room depth) to the vertical distance between the top of daylight wall to the bench surface. In labs with high requirements in precision, the depth factor should be smaller than 2; in labs with lower requirements in precision and normal offices, it should be smaller than 3.5~4. If the room is illuminated by two sides, the depth factor could be doubled.

• The labs with requirements for consistent temperature and humidity are suitable for day light, equipped with additional local lighting. This could save large amount of energy. Compared with labs with artificial illumination and no windows, this type of lab works better for people's working habits. The daylight windows shouldn't be too large and they should be well distributed to avoid oversize conduction loss of energy. When the temperature tolerance is lager than ±1~2°C double glazing could be adopted. If the temperature tolerance is lower than ±0.5°C or the environment contamination is heavy, daylight is not suggested.

③ Emergency Lighting

• The lab building should be equipped with emergency power. Once the common power is cut, it could ensure the lighting of evacuation exits and critical places and the power requirements of emergency facilities. To avoid the dangers in the experiment process due to power cut, the standby generator should ensure the power and provide the lighting within 45 seconds (the illumination could be 1/8 of the normal illumination). The emergency lighting and evacuation pilot lamps of evacuation exits, emergency exits and stairs can use storage battery power supply and the illumination should not be lower than 1 lux.

• Around the equipments which may malfunction, the walls along control room and evacuation exit, the top of the emergency exit, the corners of staircase and hallway, emergency lighting should be highlighted with striking colourful signs. The height of the light should not be higher than the height of sighting line. Foot lights are suggested to be installed in the hallway with stairs. Emergency lighting should avoid using fluorescent lights or high pressure mercury lamps which launches slow.

• In emergency exits and evacuation exits, safety evacuation lights should be installed, so that people could evacuate quickly when the emergency occurs.

(5)The Storage and Treatment of Chemical Reagent

• In new construction and retrofits of lab buildings, designers should consider to accommodate a dedicated store room and some dedicated process equipments and exhaust equipments.

• The store room of chemical reagent should meet the safety requirements, equipped with necessary sub-package, distribution and transportation tools, to ensure safety in use and collection convenience.

• The lab should have small storage cabinets to store chemical reagents used in the short time. The cabinet should be well fire-resistant and ventilated, equipped with necessary safety equipments, to ensure the safety of the reagents and hazardous substances.

• In the stores of inflammable, explosive or hazardous gas, gas mask (respirator) and specific fire extinction equipments should be equipped in the locations easy to take.

• For some dangerous work, if experimenters can't watch it all the time or can't always watch it, a dedicated high-risk laboratory should be accommodated. It should be built in a single building (outside the lab building), equipped with automatic fire-proofing, anti-explosive devices and automatic checkout equipments, with a direct exit to the outside.

(6) Accident Alarm System

The necessary accident alarm devices that a R & D institution should configure are typically as follows:

• Long-distance alarm system
• Outdoor accident phone
• Accident location display screen
• Exhaust system automatic fire alarm device (consisting of fire detector and alarm display device)

6. Construction Criteria

The unit costs for laboratory buildings are based on the construction quality and design features in the table 1. This information has been generally organised under uniformat headings. The items marked with indicate features required by government mandate for which there is "no market comparable."

Substructure Foundation	
Standard Foundation	• Allowable soil bearing pressure of 2 tons/SF assumed for spread footings • Reinforced concrete spread footing 80 PSF concrete and 2.5 PSF reinforcing
Substructure Envelope	
Basement Excavation	• 16'- 0" excavated one subgrade floor level and elevator pits
Basement Walls	• 16'- 0" (h) by 1'- 0" (thick) reinforced concrete wall resting on spread footings
Slab on Grade	• 4000 PSI 6" concrete slab with welded heavy wire mesh (20-25" fly ash) • Moisture barrier • Gravel base and compacted fill • Sealant at joints and wall junctures
Shell Superstructure	

Floor Construction	• Poured-in-place reinforced concrete floor slab • The laboratory floor framing consists of one-way concrete joists and concrete beams on the column lines • Joists are 24½" (d) with 6" (w) ribs spaced 36" OC; beams are 24½ (d) • Volume of concrete for columns is 1 cubic yard per 400 SF floor area • Column reinforcement is 1.0 PSF floor area • Total tonnage of reinforcement for the floor (slabs + beams) is 8.0 PSF • Spandrel beams along long sides of the building with column spacing 20'-0" are 24½" (d) with additional 30 lbs. per linear foot of reinforcing at all floors • Spandrel beams along short sides of building with column spacing of 30'-0" are 36" (d) with additional 40 lbs. per linear foot of reinforcing at all floors • 150 lbs/SF floor load at lab modules • 150 lbs/SF storage and receiving areas
Roof Construction	• Poured-in-place reinforced concrete frame with joist slab
Lifts	• 6" CMU shaft
Fire Egress Stairs	• 6" CMU shaft
Shell Exterior Closure	
Exterior Wall	• Insulated metal panel curtain wall system on metal stud backup • Metal stud backup with 5/8" GWB on inside • Batt insulation in metal stud cavity with vapor barrier on cold side • 24" parapet with stainless steel cap and metal stud backup wall
Corner Stone	• Cast stone
Exterior Glazing	
Fenestration	• 40% glazing/60% skin (for all)
Curtain Wall System	• Aluminum framing with 3-coat baked painted finish • Glass to be insulated double glazed units with annealed coated low-e glass; U-factor for glazing = 0.32; shading coefficient for glazing = 0.35

Window System	• Aluminum frame punched window system • Glass to be insulated double glazed units with annealed coated low-e glass; U-factor for glazing = 0.32 shading coefficient for glazing = 0.35 • Sill at 30" above floor
Exterior Doors	
Entrance Vestibule	• Double set of automatic sliding doors including track, operator, jamb and door panels • Overhead concealed electrical linear operator • 7"- 0" (w) by 7'- 0" (h) • Sliding panel to be aluminum frame glass panel with intermediate rail; door panel to swing out 90 degrees for emergency egress • Glass to be safety tempered glass • Provide keyed lock with panic release and automatic access control via card reader system
Emergency Egress Doors	• Hollow metal 1¾" insulated door 3'- 0" (w) by 7'- 0" (h) • 16 gauge steel frame with thermal break • Keyed lever lockset with panic release bar on inside and automatic access control via card reader system • Automatic closers
Fire Doors	• Overhead coiling fire doors • Concealed overhead installation • 20 gauge metal interlocking slats • Nylon smoke seals • Visual and audio (strobe) announciator to warn of operation
Coiling Overhead Dock Doors	• Concealed overhead coiling door • 26 gauge flat metal slats • Motor operation • Bottom lock • Weather seals at the bottom, guides, and hood
Vents and Areaways	• Architectural drainable steel louvers with 6" (d) adjustable blades with rain gutter

Shell Enclosure Roof	·
Roof Covering	• EPDM, single-ply membrane roofing system • Gravel ballast
Insulation	• Two layers 2" (thick) closed cell polystyrene rigid insulation
Roof Access	• Interior permanent stair extending up from emergency egress stairs with standard exterior metal door
Smoke Hatch	• 14 gauge painted steel hatch and curb unit
Interior Construction	
Partitions	
Entrance Vestibule, Public Lobby and Exit Corridors, Tenant Demising Partitions, Public Toilets, Security Office, Vending/Concession Areas, Building Maintenance, Loading Dock, Mail Room	• Structural slab-to-slab • 5/8" GWB on metal studs at 24" OC • Acoustical insulation filling the GWB wall cavity
Mechanical and Electrical Equipment Rooms	• Structural slab-to-slab • 1 hr fire rated • 55 STC • Two layers ½" GWB both sides on metal studs 16" OC • Acoustical insulation filling the wall cavity
Fire Command, Janitor Closets, Electrical Closets, Telephone Closet, Trash Room, General Storage	• Structural slab-to-slab • 5/8" GWB on metal studs at 24" OC
Ventilation, Plumbing, and Vertical Backbone Shafts	• 2 hr fire rated • 50 STC • Type X GWB shaft wall system with one layer 1" channel mounted GWB and one layer ½" GWB outside face

Emergency Egress Stairs and Elevator Shaft	• 6" CMU with and one layer ½" GWB on metal furring outside face
Public Toilets, Security Office, Vending/ConcessionArea, Fire Command, Janitor Closets, Electrical Closets, Telephone Closets	• Solid core 1¾" hardwood veneer doors 3'- 0" (w) by 7'- 0" (h) • Doorframes will be a minimum 14 gauge metal frame construction • Hardware to be locksets with levers • Key locks
Building Maintenance, Loading Dock, Mail Room, Trash Room, and General Storage	• 1" ABS plastic clad wood core double service doors 5'- 0" (w) by 7'- 0" (h) • 250 degree cam hinge system • Acrylic view window • Impact plates and cart bumpers
Mechanical and Electrical Equipment Rooms	• Hollow metal 1¾" double doors 6'- 0" (w) by 7'- 0" (h) • 16 gauge welded metal frames • Hardware to be locksets with levers • Key locks
Emergency Egress Stair Doors	• Fire-rated solid core 1¾" hardwood veneer doors 3'- 0"(w) by 7'- 0" (h) • 16 gauge welded metal frames • Hardware to be panic release with levers opposite side • Automatic closers
Specialties	
Specialties – Handrail	
Emergency Egress Stairs	• Welded pipe handrail
Specialties – Toilet Accessories	• Stainless steel ceiling hung partitions • Toilet paper holder • Feminine napkin disposal (female toilets only)

	• Feminine napkin dispenser (female toilets only) • Paper towel dispenser combination waste receptacle • Soap dispenser • Mirror with stainless steel edging • ADAAG compliant toilet grab bars
Fire Extinguisher Cabinets	• Fire extinguisher cabinets in storage rooms and equipment rooms
Signage	
Building Directory	• Touch-screen computer monitor programmed building directory • Stone veneer pedestal case
Dedication Plaque	• Bronze 4 SF with raised letters
Floor Identification	• Dimensional letters mounted on wall covering with ADAAG compliant tactile Braille signage
Emergency Egress	• Etched on plastic laminate signage system panel with ADAAG compliant tactile Braille signage
Room Identification for Major Public Spaces	• Room identification signage to be raised plastic letters mounted beside the door with ADAAG compliant tactile Braille signage
Room Identification	Signage system to be modular vinyl lettering on plastic laminate signage frame system with ADAAG compliant tactile Braille vinyl signage modules
Telephone Enclosure	• Stainless steel dividers with stainless steel shelf and perforated interior face with acoustical material
Interior Finishes	
Walls	
Main Lobby, Main Lift Lobby	• 5'- 0" (h) stone wainscot with Type II vinyl wall covering above
Upper Floor Lift Lobby, Public Corridors	• Type II vinyl wall covering with hardwood base
Public Toilets	• 3/8" textured porcelain tile base and wainscot with paint above
Vending/Concession Area, Copier Area	• Low VOC paint with vinyl cove base
Security Office, Egress Corridors	• Low VOC paint with vinyl cove base

Building Maintenance, Loading Dock, Mail Room, Trash Room, General Storage	• Low VOC paint with vinyl cove base • Vinyl chair rail guard and vinyl corner guards
Mechanical and Electrical Equipment Rooms, Fire Command, Janitor Closets, Electrical Closets, Telephone Closet	• Low VOC paint with vinyl cove base • Steel corner guards
Floors	
Entrance Vestibule	• Entrance to have 1" terrazzo floor tile 12" by 12" with mastic base • Drained entrance grid with structural aluminum rails, drain pan, and carpet tread inserts of monofilament solution died nylon fusion bonded to backing
Main Lobby, Main Lift Lobby	• Terrazzo tile
Upper Floor Lift Lobby	• Terrazzo tile
Public Corridors, Security Office	• Broadloom carpet • 32 oz face weight • Yarn dyed colour • Fourth generation nylon yarn • Bonded construction with cushioned back
Public Toilets	• 3/8" textured porcelain tile
Vending/Concession Area, Copier Area	• Vinyl composition tile
Building Maintenance, Mail Room, Trash Room, General Storage, Janitor Closets, Fire Command	• Vinyl composition tile
Loading Dock, Mechanical and Electrical Equipment Rooms	• Sealed concrete

Electrical Closets, Telephone Closets	• Vinyl composition tile
Ceiling	
General	• Suspended 24" (w) by 24" (l) acoustical tile ceiling
Entrance Vestibule	• Plaster
Main Lobby, Main Lift Lobby, Upper Floor Lift Lobby	• Low VOC painted GWB
Upper Floor Public Corridors	• Suspended 24" (w) by 24" (l) acoustical tile ceiling
Public Toilets	• Suspended 24" (w) by 24" (l) acoustical tile ceiling • Soffit over counter areas
Vending/Concession Area, Security Office	• Suspended 24" (w) by 24" (l) acoustical tile ceiling
Egress Corridors	• Suspended 24" (w) by 24" (l) acoustical tile ceiling
Building Maintenance Office, Mail Room, Fire Command	• Suspended 24" (w) by 24" (l) acoustical tile ceiling
Building Maintenance, Shop Area, Trash Room, General Storage, Loading Dock, Mechanical and Electrical Equipment Rooms, Janitor Closets, Electrical Closets, Telephone Closets	• Exposed structure above
Conveying Systems	
Lifts	
Public Lifts	• Holed hydraulic lift • Lift cab allowance: $31,500/per cab (Oct '00 dollars)
Service Lifts	• Holed hydraulic lift • Lift cab allowance: $5,000/per cab (Oct '00 dollars)

Plumbing	• Vinyl composition tile
Utility Service: Domestic Water Supply	• Two domestic cold water services shall be provided connecting to the public utilities in the adjacent streets • Domestic cold water services shall be metered in accordance with local requirements • Domestic water services shall be equipped with reduced pressure type backflow preventors located on the first level above grade
Utility Service: Storm Drainage and Sewerage Systems	• Multiple sanitary and storm water (primary and secondary) house drain services shall be provided from the building and connect to public utilities in adjacent streets
Utility Service: Natural Gas	• A natural gas service shall be extended into the building and be metered in accordance with local requirements • Shut-off valve at gas service entry point
Public Toilets	• Porcelain sink inset in counter • Cold and hot water supply • Lever faucet • Porcelain floor mounted flush-valve water closet • Floor drain with primer
Domestic Cold Water System	• Each system shall be pressurised by a factory prefabricated tri-plex constant pressure pumping system • Provide independent domestic cold water systems for general building and for laboratories • All domestic water connections to non-potable sources shall be provided with suitable backflow preventors • Provide non-freeze hydrants around the base of the building located on each side of main entrance and spaced approximately 100'-0" OC around building
Domestic Hot Water System	• Provide independent domestic hot water systems for general building and for laboratories • Domestic hot water for each system shall be generated by multiple gas-fired storage type water heaters with water heater flues to be extended through main roof • A multi-zone central domestic hot water distribution system with supply and recirculation piping shall be provided to serve all fixtures and equipment requiring hot water; recirculation shall be provided to any fixture located greater than fifty feet from a circulated main or riser

Sanitary Drainage Systems	• All areas below grade shall be provided with duplex sewage ejector stations; each ejector pump shall be sized for 100% capacity and be provided with emergency power
Vending/Concession Area	• Cold water supply with shut-off at connection
Drinking Fountains	• Wall-mounted fountain with chiller
Laboratory Utility Systems	• Six nominal piping systems • Central distribution for distilled water system with high purity dual bed deionised water feed with polypropylene piping • Central distribution for central natural gas supply with medical grade copper tubing • Central distribution and service for special gas supply systems including O2, N2, He, CO2 with medical grade copper tubing • Central distribution, pumps and reservoir for central vacuum of 18"- 22" of mercury in medical grade copper tubing; discharge to have solvent stripper recovery unit followed by carbon filters • Central collection risers and mains for acid waste and vent with polypropylene piping with treatment system consisting of limestone chip acid neutralisation tank discharged to the sanitary sewer system • Central collection risers and mains for solvent waste and vent with glass piping passed through solvent stripper/recovery unit of vacuum condensing type with waste discharged to acid waste system
Mechanical Room, UPS Battery Rooms	• Floor drain with primer • Emergency eye wash and deluge shower
HVAC	
General	• All HVAC systems and equipment shall at minimum comply with the energy performance criteria within the "Facilities Standards for the Public Buildings Service" supporting an assigned energy performance goal • System and equipment selections indicated below are for the purposes of this unit cost study only; alternate system and equipment options should be investigated on a specific building project for improved efficiency of operation, and enhanced life cycle economic performance
Design Conditions and Loads	• Outdoor design conditions shall be as per GSA Standards • Indoor design conditions shall be as required • Ventilation rates shall meet or exceed all required codes and standards, including ASHRAE-62, but in no case be less than 20 CFM of outside air per occupant • Space-heating boilers have been sized assuming a design load of 30 Btu/h per GSF of building • Central cooling equipment has been sized on the basis of 1 ton of refrigeration per 300 GSF for unit cost purposes; However, designers shall minimize cooling capacity to the degree possible while also satisfying all design criteria

Energy Supply	• A complete fuel oil pumping system shall be provided for the emergency generators and boilers and shall include fuel oil storage tanks, piping, valves, duplex fuel oil pump and day tank • Tanks to be buried underground double-walled fiberglass tanks with leak detection system • See Plumbing – Utility Service: Natural Gas for criteria
Heat Generating System	• Heating system shall be hot water type generated by dual fuel boilers (natural gas and #2 fuel oil); provide oil storage tank • Hot water shall be distributed to perimeter fan coil units and perimeter fan powered VAV boxes with heating coil • Heating water shall be distributed by two hot water pumps through two pipe reverse return system; hot water to glycol heat exchanger with two pumps (one standby) to be provided • For unit cost purposes, two space-heating boilers are assumed with each rated at approximately 67% of peak heating load; boiler capacities used in this study are as follows: two boilers at 65 HP each {capacities shown are in BHP (boiler horsepower where 1 BHP = 33,475 Btu/h)} • Pumps to be horizontal split case • Provide mechanical seals for all water pumps
Cooling Generating Systems	• Refrigeration machines shall be electrically driven chillers • For unit cost purposes, chillers are sized for 50%, 50%, and 20% of the peak cooling load; chiller capacities used in this study are as follows: 2 at 175 Tons and 1 at 70 Tons • Plate-and-frame heat exchanger provided for free-cooling application • Cooling towers to be forced draft type steel frame with fireproof fill
Air Distribution System	
Air Supply, VAV	• Variable air volume (VAV) terminal reheat system with pre-filters and after-filters for 95% efficiency with terminal humidifier and with make up air handling unit with pre-heat and prefilter unit • Laboratory modules to have positive pressure relative to other spaces; no return air from laboratories to other spaces
Air Handling Units	• Provide a minimum of one unit for every floor and a separate unit for every 25,000 CFM of capacity • The air handling system(s) to consist of recirculating variable air volume air conditioning units providing conditioned air on each floor for space cooling and ventilation; each unit to consist of a supply air fan, filters, chilled water coil, sound attenuation and controls • Fan motors shall be driven by Variable Frequency Drives (VFD) for efficient electrical operation

	• Minimum outside air for each fan room will be supplied from a central outside air fan system which includes filters, cooling coil, heating coil and humidifier
Materials	• Sheet metal work for gauges and bracing shall conform to ASHRAE and SMACNA standards • Pipe for chilled water, condenser water, steam and hot water piping to be schedule 40 standard with steel ASTM A53 lap welded or seamless black steel • Valves to be furnished and installed as necessary for the control and easy maintenance of all piping and equipment • Expansion loops to be provided for all piping systems • Grilles, registers and diffusers to be provided as required • Dampers to be provided as required for proper balancing of systems and all fire and fire/smoke dampers required by code • Fans (centrifugal) to be airfoil type with adjustable sheaves below 50 HP • Air filters to be 25-30% efficiency prefilters and 80-85% efficiency final filters for each AHU • Insulation for sheet metals to be provided in all medium pressure supply air ductwork from fan discharge to pressure reducing device (including flexible connections) and low-pressure ductwork; all supply, return, spill, outside air intake and exhaust plenums to be insulated
Exhaust Air	• Toilets to be provided with 100% exhaust operated by time clock or building management system • Ducted ceiling exhaust ducts with economiser exhaust and connections to individual fume hoods • Provide ducted exhaust system from laboratory modules for fume hood and safety cabinet exhausts • Provide dry trap for condensing solid material, one bubbler for reduction of post-reactions initiated by the presence of oxygen with caustic scrubbing liquid, a nitrogen purge, and a charcoal trap to remove un-reacted toxic gases • Provide flexible exhaust duct "snorkel" connection — 1 per laboratory module • Emergency generator vertical exhaust • UPS battery room to have 100% direct exhaust
Controls	• Building Automation Systems: all building systems shall be monitored or controlled or interfaced through the Building Automation System (BAS) which BAS consists of an Energy Management System (EMS), Security System and Fire Protection System; system selection shall be expandable and allow communication with other automation systems • The EMS will have Central Processing Unit (CPU), monitor, local permanently mounted alphanumeric keyboard, printer, control, and feedback functions; software programmes will be

	used for control; all systems will be provided with redundant backup • The EMS shall utilise Direct Digital Controls (DDC) for system control; monitoring the systems will be accomplished with a central terminal in the BAS office; control systems shall be pneumatically actuated • Alarm to be the BAS system shall notify the operator of equipment failures and high/low operating conditions in all systems
Fire Protection	
Service	• Two services connecting to public utilities in adjacent streets • Fully metred in accordance with local requirements • Equipped with reduced pressure type backflow preventors located on the first level above grade
Fire Suppression	• Combination fire standpipe/sprinkler system throughout the building pressurised by automatic electric fire pump and jockey pump • Fire pump shall be supplied with normal and emergency power and an automatic transfer switch • Automatic wet pipe sprinkler system throughout except areas subject to freezing where a dry pipe system shall be used • Recessed automatic glass bulb quick response type sprinkler heads; provide one sprinkler head for every 100 SF of finished space • Elevator machine room, elevator shafts and electrical switchgear rooms with sprinkler systems; cooling towers with deluge type sprinkler system • Fire department hose valves at stairways shall consist of a hose valve within the stair and an additional valve on the corridor side of the stairwell • Siamese connections • Tamper switches on all fire protection control valves • Each sprinkler floor system connection to standpipe riser and main provided with OS&Y gate valve with tamper switch, check valve, water flow alarm, inspectors test and drain, drain with sight glass • Multipurpose ABC dry chemical fire extinguisher in storage rooms and equipment rooms
Fire Alarm System	• Addressable type, electronic fully supervised multiplexing type employing high frequency carrier applied to dedicated wires for the distribution of its multiplex coded signals • Fire safety system command centre in room on lobby level with direct access for fire fighters; command centre to receive local alarms; remote annunciator panels located in engineer's control room • Fire protection alarm system devices shall be located in accordance with the following: manual fire alarm pull station adjacent to exit door on each floor; space smoke detectors (analog type) in all elevator lobbies, electrical switchgear, transformer vaults, and telephone exchanges; intercom (warden) stations on each floor and in each mechanical room; duct smoke detectors (analog type) in air handling systems in excess of 2,000 CFM; water-flow detectors in sprinkler

	piping; tamper switches on valves in sprinkler piping; automatic control (stopping) of air handling systems in response to signal from the fire protective alarm system and automatic starting of smoke exhaust and pressurisation fan systems; manual control of fans from the fire command centre; combination voice evacuation speaker and visual devices throughout the floors, visual device in each toilet; elevator recall to ground floor
Smoke Evacuation	• Ceiling hatches in stairwells • Automatic opening ventilation louvers at stairwell bases • System actuated ventilation fans
Electrical	
Electrical Service	• Suitable for receiving low-tension power at the 480/277 volt level from facilities provided by the utility company
Service and Distribution Equipment	• Include all the elements necessary to conduct electricity in an approved safe manner to all lighting fixtures, air conditioning equipment, heating equipment, plumbing equipment, sanitary equipment, elevators, special electrical systems, receptacle and appliance outlets, and signal and communications equipment • Single supply connection main switchboards • All required subsidiary panelboards (power, distribution, lighting, and appliance) • Automatic power factor correction equipment for each switchboard to maintain a 90% power factor • Incorporate copper busses and copper wiring throughout • 480 volts, three phase for all motors ½ horsepower and larger • 277 volts single phase to all fluorescent (and other discharge type lamp) lighting fixtures • Power conditioning and transient suppression (PCTS) devices for each main switchboard, main emergency distribution panelboard, and each 120/208 appliance panelboard • Three phase dry type 115° C transformers (480-120/208) for all normal power requirements • Three phase dry type K-13 rated transformers (480-120/208) for all panelboards serving office automation equipment and work stations • 120/208 volt appliance panelboards serving office automation (electronic) equipment shall be suitable for "harmonic rich" line to neutral loads • Grounding to consist of a series of driven ground rods and cable with connections to grounding electrodes • Provide master labeled UL96 lightning protection system • Plug-in buss duct risers will be utilized for distributing normal power to each of the floors
Emergency Power	

Generator Unit	• Diesel-driven emergency generator unit with paralleling switchgears for multiple generators; provide 500 KW unit • Automatic transfer switches (by-pass isolation type) arranged to maintain the emergency power distribution system energised from the normal utility company source or the generating set • Remote emergency alarm panel for each generator located at the building control centre
Uninterruptible Power Systems	• Provide separate uninterruptible power systems complete with UPS modules with 30-minute battery backup, maintenance bypass switchgear, and interconnecting circuitry for the computer/data and communications systems, life safety lighting systems, and security systems
Electrical Outlets	
General Areas	• Wall mounted duplex outlets every 50'- 0" OC
Corridors and Lobby Spaces	• Wall mounted duplex outlets every 50'- 0" OC • Provide a dedicated line duplex electrical outlet at the public lobby for metal detector and X-ray security screening equipment • Provide recessed duplex wall receptacle for clock in each lobby and corridor
Vending/Concession Area	• One quadplex counter splash mounted electrical outlet • One duplex wall outlet for each vending machine
Electrical and Communication Closets	• Two dedicated duplex outlets on emergency power plus additional outlets for every 6'- 0" of wall space • A separate 120-volt panel with master switch and five 20-amp circuits to be included for each telephone and LAN system for each separate agency
Maintenance Shop, Mail Room	• Wall mounted duplex outlets every 50'- 0" OC
Public Toilets	• Ground fault electrical duplex outlet
Lighting	
Entry Vestibule	• Recessed down lamps one per every 10 SF
Main Lobby, Main Elevator Lobby, Upper Floor Elevator Lobby	• Metal halide uplighting
Public Corridors, Egress Corridors	• Parabolic fluorescent 24" (w) by 48" (l) recessed ceiling fixtures with two T-8 lamps and electronic ballasts located every 80 SF (or T-5 equivalent)
Public Toilets	• Recessed fluorescent light fixture located in the soffit above the lavatory and the toilet

Vending/Concession Area, Security Office	• Parabolic fluorescent 24" (w) by 48" (l) recessed ceiling fixtures with two T-8 lamps and electronic ballasts located every 80 SF (or T-5 equivalent)
Building Maintenance Office, Mail Room, Fire Command	• Parabolic fluorescent 24" (w) by 48" (l) recessed ceiling fixtures with two T-8 lamps and electronic ballasts located every 80 SF (or T-5 equivalent)
Building Maintenance Shop Area, Trash Room, General Storage, Loading Dock, Mechanical and Electrical Room, Janitor Closets, Electrical Closets, Telephone Closets	• Suspended fluorescent 24" (w) by 48" (l) recessed ceiling fixtures with two T-8 lamps and electronic ballasts located every 80 SF (or T-5 equivalent)
Telephone and Communication Outlets	• Conduit, power, and mounting/telephone boards for telephone and data communications system are provided as part of the building shell and core unit costs; equipment and wiring provided by tenant
Public Lobby	• Conduit and boxes for telephone connections for security screening post provided as part of the building shell and core unit costs; equipment and wiring provided by tenant • Conduit and boxes for public pay telephone connections provided as part of the building shell and core unit costs; equipment and wiring provided by tenant • Conduit and boxes for one data connection for electronic building directory provided as part of the building shell and core unit costs; equipment and wiring provided by tenant
Security Office, Building Maintenance Office, Mail Room	• Conduit and boxes for one telephone line provided as part of the building shell and core unit costs; equipment and wiring provided by tenant • Conduit and boxes for one LAN connection provided as part of the building shell and core unit costs; equipment and wiring provided by tenant
Telephone Room	• Four 4" vertical conduits between floors provided as part of the building shell and core unit costs • Conduit and boxes for one telephone line provided as part of the building shell and core unit costs; equipment and wiring provided by tenant • Conduit and boxes for mounting board for telephone and LAN switch connections provided as part of the building shell and core unit costs; equipment and wiring provided by tenant
Mechanical Room	• Conduit and boxes for one telephone line provided as part of the building shell and core unit costs; equipment and wiring provided by tenant • Conduit and boxes for one LAN connection for BAS computer provided as part of the building shell and core unit costs; equipment and wiring provided by tenant
Security Devices	

General	• Exterior intrusion detection system, including door position detectors and lock keeper detectors on all exterior doors, glass break sensors on all exterior glazing, and volumetric motion sensors outside each door • For interior security, conduit, power and mounting support for interior security devices including X-ray baggage and walk through metal detectors provided as part of the building shell and core unit costs
Entry Vestibule, Entry Door from Restricted Parking, Dock Man Door and Cargo Overhead Door	• Card reader access control system • Intrusion detection system, with door position detector and lock keeper detector and glass break sensors • Intercom and duress alarm • Closed circuit television monitor • Volumetric motion sensor
Emergency Egress Doors	• Intrusion detection system with door position detector and lock keeper detector • Glass break sensors • Closed circuit television monitor
Building Perimeter	• Glass break sensors • Closed circuit television monitor
Public Lobby	• Closed circuit television monitor • Glass break sensor • Metal detector • X-ray baggage inspection equipment
Security Office	• Monitors for intrusion detection systems, duress alarms, intercoms, closed circuit television cameras, fire alarms, and card access controls
Mail Room	• X-ray package inspection system
Lift	• Remote floor recall override to stop lifts on floors away from threats
Commercial Equipment	
Window Washing Equipment	• Davit only • Allowance: $15,000 (Oct '00 dollars)

Dock Loading Equipment	• One dock leveler with electro-hydraulic operation for building services provided as part of the building shell and core unit costs; additional docks will be tenant assignable space and associated levelers and dock equipment will be a special cost to the tenant
Furnishings	
Casework	
General	• All millwork to be AWI custom grade plastic laminate veneer panels with stainless steel
Public Toilets	• Cantilevered plastic laminate counter with splash
Public Lobby, Security /Information Desk	• Wood veneer construction with transaction surface of polished granite and worksurface of plastic laminate
Vending/Concession Area, Security Offic	• Painted metal base and upper cabinets • Plastic laminate counter with splash
Building Site Work	
General	• Site work allowance carried in estimate to cover such items as: roadways, walkways and plazas, vegetation, site lighting, and site utilities • Site allowance is based on a site area to GSF ratio of 75%
Flagpoles	• 30'- 0" (h) aluminium pole with internal halyard and spread footing base
Roadways	• Concrete 12'- 0" (w) lanes with curbs
Walkways and Plazas	• Concrete walkways
Fountains	• Round fountain in entrance plaza
Vegetation	• Grass ground cover • Accent annual flowerbeds and flowering shrubs along entrance paths • Perimeter indigenous trees
Site Lighting	• Metal halide high mast general lighting • Metal halide building security flood lighting

Cases of Laboratory Space Planning

1

2

3

The function is reflected in the exterior form and patterns

| John Curtin School of Medical Research,
The Australian National University

Location:
Canberra, Australia
Architect:
Lyons
Construction Area:
7,700 m²
Completion Date:
2007
Photographer:
John Gollings Photography

▪ The function is reflected in the exterior form and patterns. Two aluminium "strands" at the top and bottom of the façade are articulated continuously around the building, which imply the 3D image of double helix structure in molecular biology.

▪ The "super-lab" laboratory modules and planned for optimum flexibility for small groups or large research teams. All laboratories include flexible services and moveable benching and have abundant natural daylight.

▪ Features of entrance hall, stairs and balcony: the wall of entrance hall is a combination of aluminium composite fins and floor-to-ceiling glass. Lifts, interconnecting stairs and balconies are located centrally in the new building. These provide a focus for circulation and social interaction.

1. The façade of the building expresses the work undertaken by the School – both formally and in the texture of its surface – using the double DNA strand (synonymous with medical research) as the generative idea.
2.3. Two aluminium "strands" at the top and bottom of the façade are articulated continuously around the building. On the entry façade, these twist around an invisible centreline, alluding to the popular three dimensional image of the double helix used in molecular biology.
4.5. Digitally formed precast concrete panels represent the work of the school at four different scales, from the scale of the human body/humankind to the cellular, the DNA molecule and the chemical bases which make up the DNA molecule represented by A, G, C and T.
6.West façade of the building

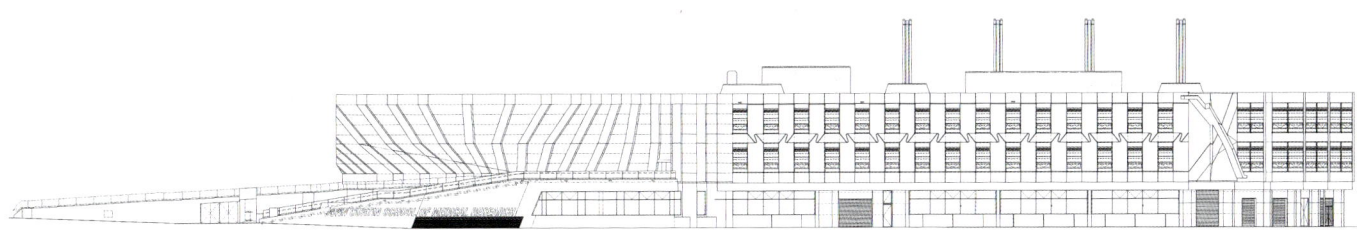

North Elevation

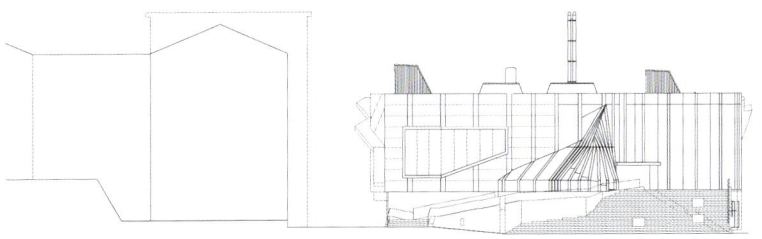

East Elevation

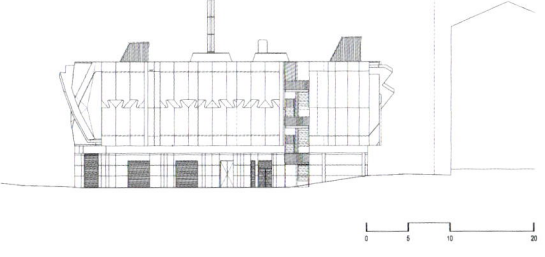

West Elevation

7

7. The foyer walls are composed of aluminium clad fins with full height glass. They provide passive control of solar penetration and frame views into the foyer space and its interactive displays.

8. Lifts, interconnecting stairs and balconies are located centrally in the new building, providing a focus for circulation and social interaction.

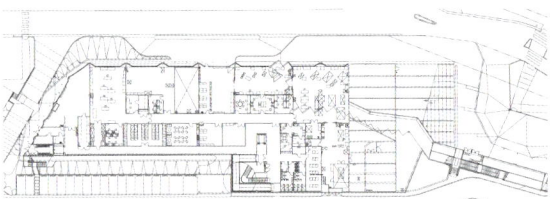

Ground Floor Plan

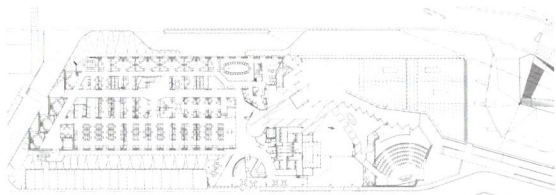

First Floor Plan

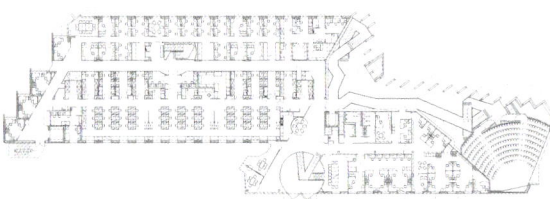

Second Floor Plan

8

TIPS:

Aluminium for recycling:

Aluminium is theoretically 100% recyclable without any loss of its natural qualities.

Recycling involves melting the scrap, a process that requires only 5% of the energy used to produce aluminium from ore, though a significant part (up to 15% of the input material) is lost as dross (ash-like oxide). The dross can undergo a further process to extract aluminium.

In Europe, aluminium experiences high rates of recycling, ranging from 42% of beverage cans, 85% of construction materials and 95% of transport vehicles.

Recycled aluminium is known as secondary aluminium, but maintains the same physical properties as primary aluminium. Secondary aluminium is produced in a wide range of formats and is employed in 80% of alloy injections. Another important use is for extrusion.

White dross from primary aluminium production and from secondary recycling operations still contains useful quantities of aluminium that can be extracted industrially. The process produces aluminium billets, together with a highly complex waste material. This waste is difficult to manage. It reacts with water, releasing a mixture of gases (including, among others, hydrogen, acetylene, and ammonia), which spontaneously ignites on contact with air; contact with damp air results in the release of copious quantities of ammonia gas. Despite these difficulties, the waste has found use as a filler in asphalt and concrete.

2

The design of architecture form serves interior function

| Research Centre for Molecular Medicine of
the Austrian Academy of Sciences

Location:

Vienna, Austria

Architect:

Kopper Architektur

Site Area:

690 m²

Completion Date:

2010

Photographer:

Bruno Klomfar

▪ The designer uses glazing wall and skylights in the exterior design. The transparent effect integrates with the artistic façade, generating a deep effect.

▪ The close connection between architectural form and indoor functional zones means that the ultimate goal of research buildings is to serve its functions.

▪ Terraced form and slanted façade provide a more comfortable and open environment for the discussion and communication areas on the top floor.

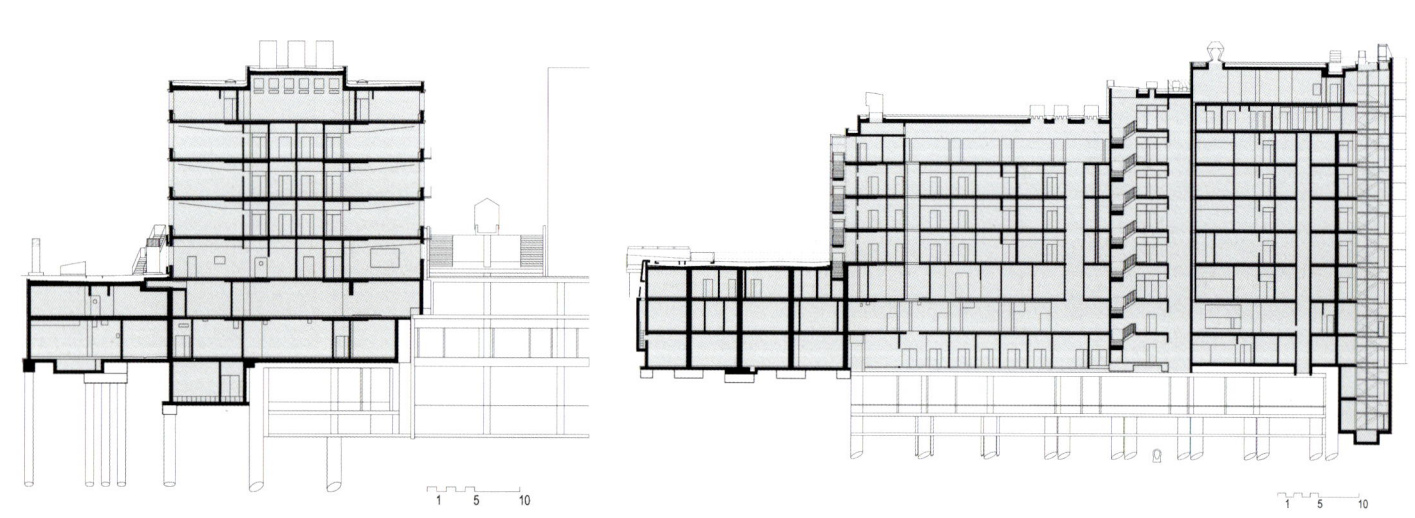

Sections

1. Main entrance
2. The original starting point was a joint lab structure based on a totally symmetrical triple-hipped formula; yet in the course of the development and of the definitions of use, the two buildings each developed in their own direction.
3. With its terraces and inclined façades, the implementation of the top facilities as a seminar and communications zone provides a clear-cut reference to the inherent functions, meaning that any formal focus remains firmly rooted in the functional aspect at all times.
4. The Eastern façade which, in architectural terms, was always designed as a membrane to which the longitudinal façades are adjacent.
5. West façade of the building
6. Office
7.8. Extensive and noticeable room and window heights, creating the impression of airy spaces
9. Conference room
10. Seminar room

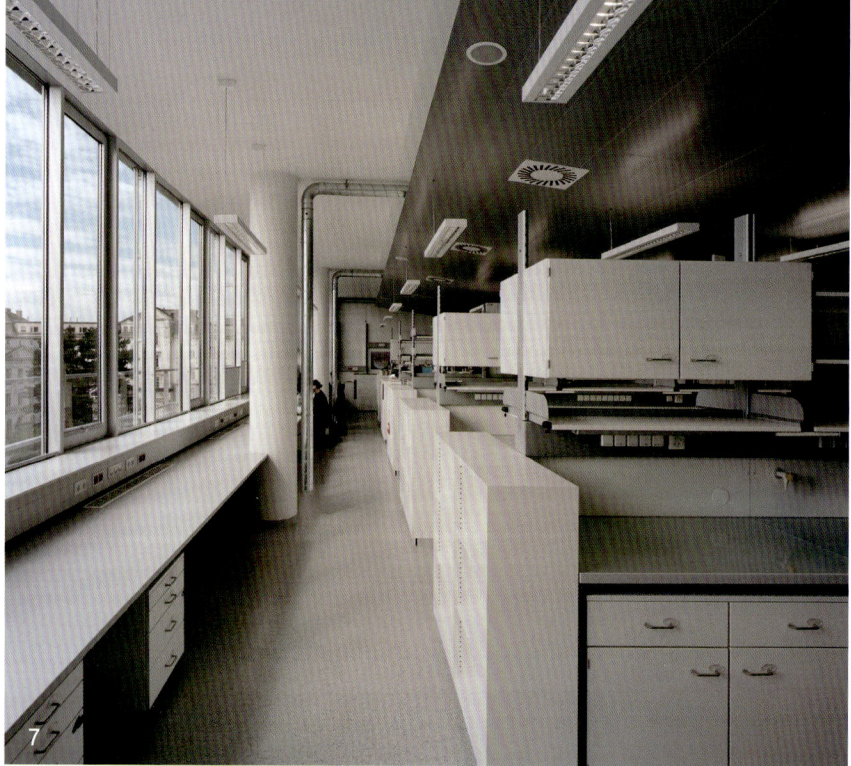

TIPS:

Spitalgasse:
The Spitalgasse is one of the streets in the Old City of Bern, the medieval city centre of Bern, Switzerland. It is part of the Äussere Neustadt which was built during the third expansion from 1344 to 1346. The eastern end is at Waisenhausplatz and Bärenplatz while the western end is at Bahnhofplatz near the Church of the Holy Ghost. It is part of the UNESCO Cultural World Heritage Site that encompasses the Old City.

Presenting the historic character, remodelling the space function

| University of Maribor, Faculty of Agriculture and Life Sciences

Location:
Maribor, Slovenia
Architect:
Styria Arhitektura
Construction Area:
7,491 m²
Completion Date:
2009
Photographer:
Miran Kambič u.d.i.a.

▪ Challenge in design: The task of changing the purpose of the use of space and the related existing building substances is classified among the most creative and responsible challenges in architecture.

▪ In planning strategy the architects intertwine two principles of designing strategies within which the old and the new form a mutual expression at different historical levels.

>>The first principle of design is the structuring of the time-, contents- and character-related components of the building on the basis of the knowledge available in the fields of archaeology, preservation,

architecture and history. The primary starting point is the presentation of the Dvor's (the Mansion's) historical image where the basic perception of aesthetics and qualitative criterion of its occurrence derive from.

>>The second principle of design is the presentation of the castle facility as a whole. The optimisation of the programme and functional expectations of the user is the starting point connected with the technology of use and with the observation of the regulations in effect in the field of the construction.

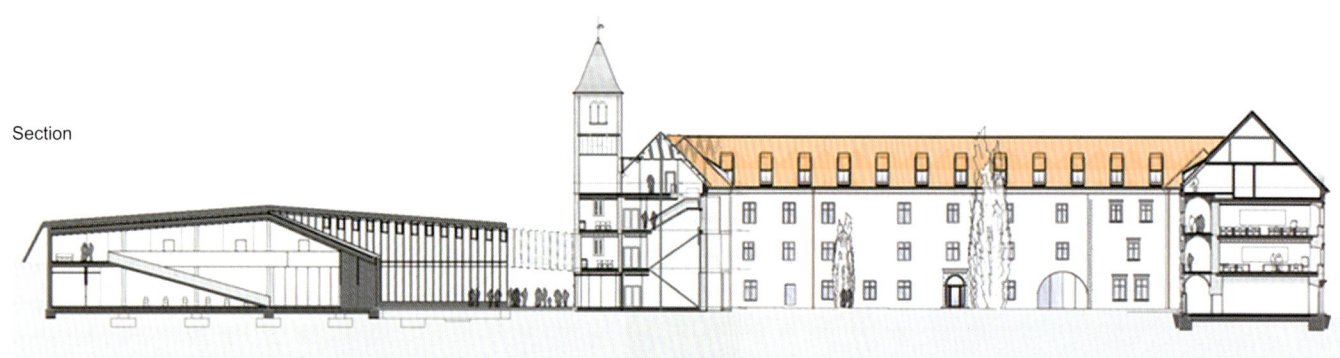

Section

1. In the shape of a cut-out landscape carpet a green meadow becomes a roof while the landscape replaces its topographic profile and is transformed into the plane of a green roof.
2. The old castle HOMPOS dated back to 11th century became a modern educational and research institution.
3. Connection with the castle over the glazed bridge
4. The auditorium extends into the external classroom through the glass wall.
5. Irregularly designed building volume calls for contact with nature.
6. Entrance hall with open gallery and natural light
7. Communication gallery with glass wall
8. Library with panoramic view to vineyards
9. Laboratories on the outside perimeter
10. Presentation of archeological relicts
11. The auditorium
12. Classroom between solid white walls

Master plan

7

TIPS:

Award name: Nominated, MIES VAN DER ROHE AWARD 2011

Award Description: European Union Prize for Contemporary Architecture Mies van der Rohe Award is granted every two years by the European Union and organised by the Fundació Mies van der Rohe, Barcelona, to acknowledge and reward quality architectural production in Europe.

In this way, the Award draws attention to the major contribution by European professionals to the development of new ideas and technologies. At the same time, it offers both individuals and public institutions an opportunity to reach a clearer understanding of the cultural role of architecture in the construction of our cities. Furthermore, the Award sets out to foster architecture in two significant ways: by stimulating greater circulation of professional architects throughout the entire European Union in response to transnational commissions and by supporting young architects as they set off on their careers.

Candidates for the Award are put forward by a broad group of independent experts from all over Europe, as well as from the architects' associations that form part of the European Council of Architects and other European national architects' associations. At each two-yearly edition, the jury selects two works: one that receives the European Union Prize for Contemporary Architecture in recognition of its conceptual, technical and constructional qualities, and the other that receives the Emerging Architect Special Mention. The jury also selects a set of finalist works to be included in both the Award catalogue and exhibition.

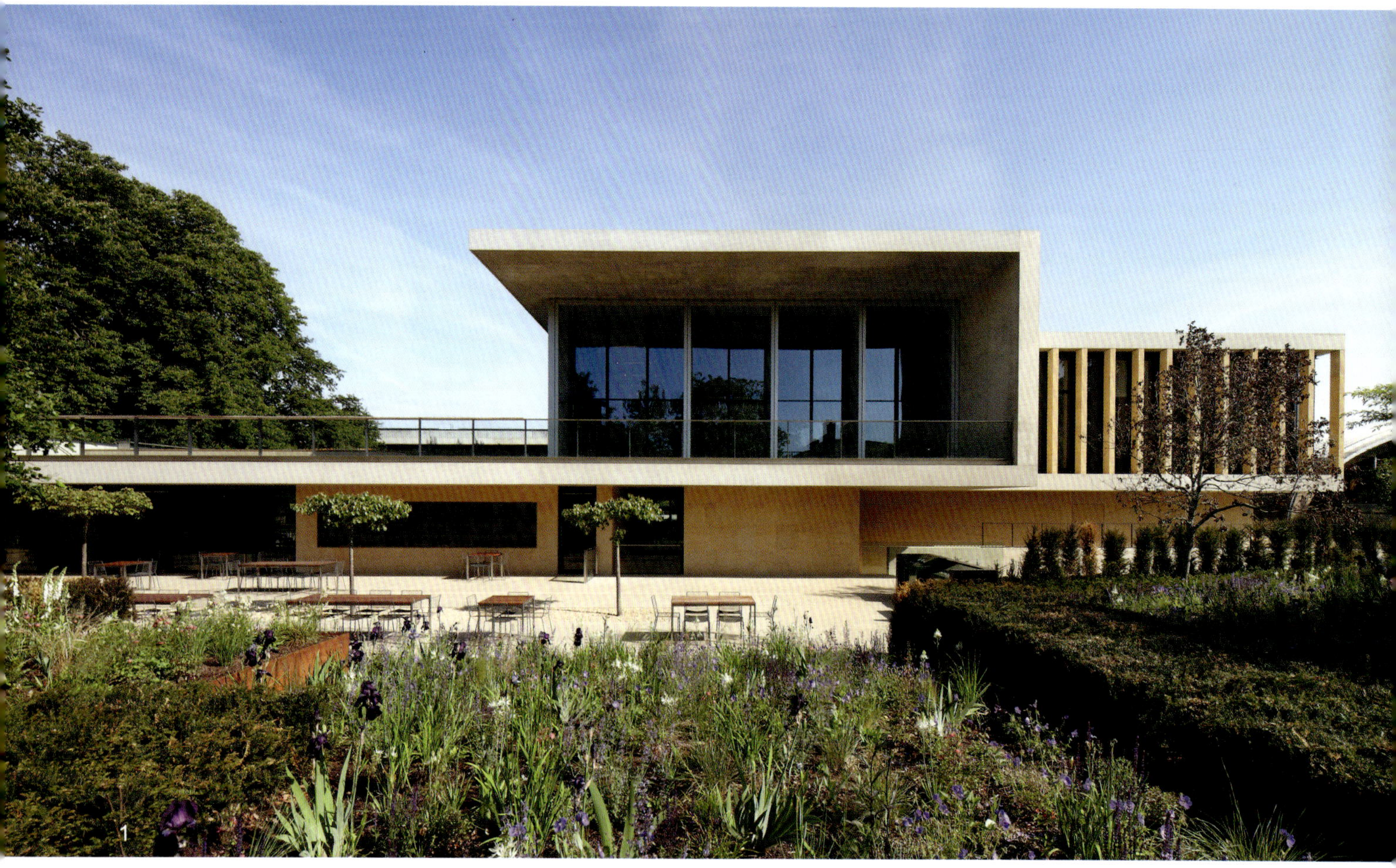
1

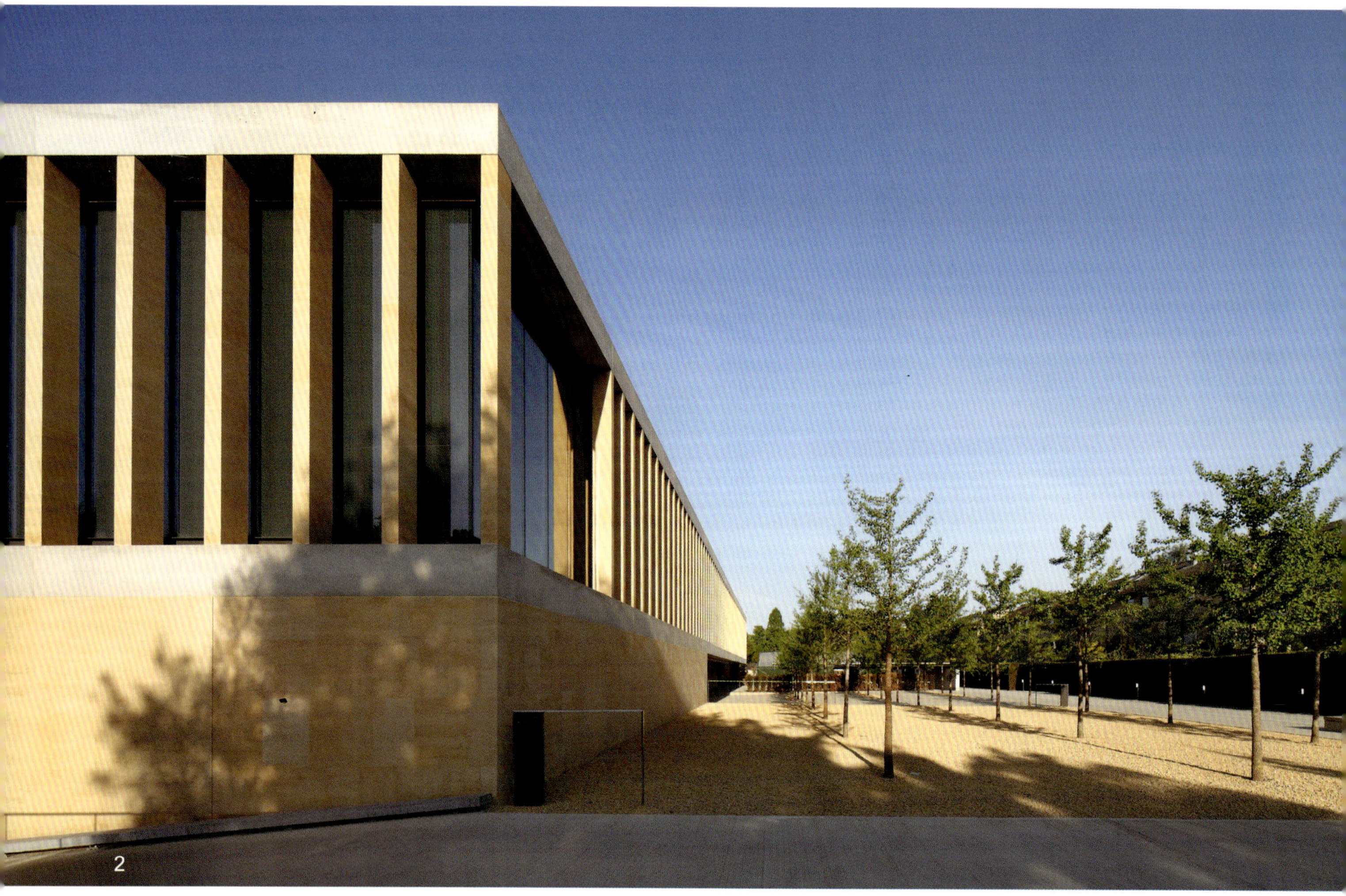
2

3

The "route" design interprets the tradition of the Greek stoa and the monastic cloister
| Sainsbury Laboratory

Location:
Cambridge, UK
Architect:
Stanton Williams Architects
Construction Area:
11,000 m²
Completion Date:
2010
Photographer:
Hufton+Crow

■ The building's identity is established externally as a series of interlinked yet distinct volumes of differing height grouped around three sides of a central courtyard.
■ The design reconciles complex scientific requirements with the need for a piece of architecture that also responds to its landscape setting. It provides a collegial, stimulating environment for innovative research and collaboration.
■ Further visual connections are created by the careful use of glazing in the building.
■ The designer uses "route" design to interpret the tradition of the Greek stoa, the monastic cloister, and the collegiate court, all of which were intended to some extent as semi-outdoor spaces for contemplation and meetings. As a result, past, present, and future are connected.

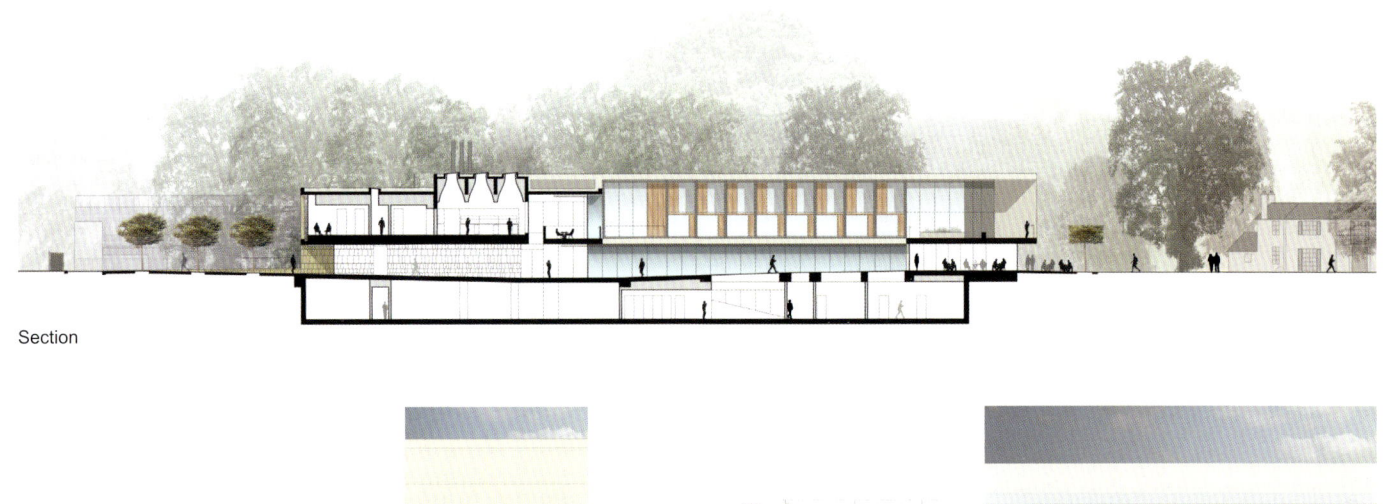
Section

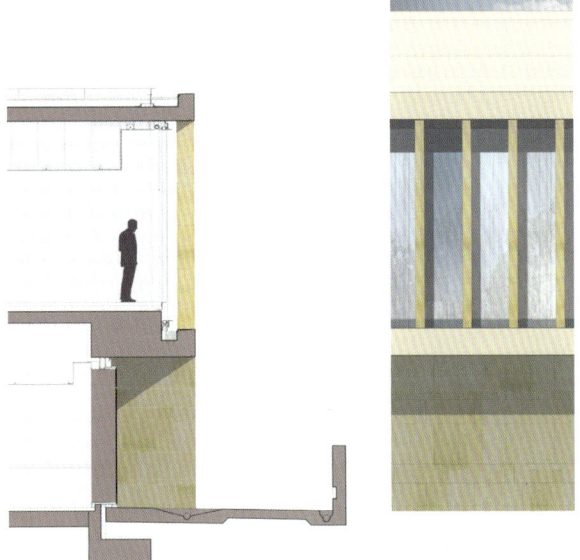
Facade Detail

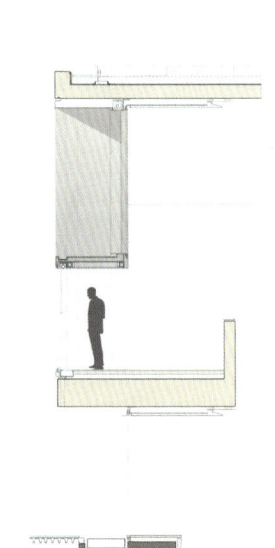

Facade Detail

4

1. Solidity is implied by the use of bands of limestone and exposed insitu concrete, recalling geological strata and indeed the Darwinian idea of evolution over time as well as the permanence which one might expect of a major research centre.

2. Its windows are screened by narrow vertical bands of stone that imbue the elevation with a regular consistency, behind which the pattern of fenestration could potentially be altered in response to future requirements.

3. The building as a whole is rooted in its setting. There are two storeys visible above ground and a further subterranean level, partly in order to ensure efficient environmental control, but also to reduce the height of the building.

4. With glazed windows facing the court on one side and internal windows offering glimpses of the laboratories on the other, the court operates as a transitional zone between the top-lit working areas at the centre of the building and the Botanic Garden itself.

5. Further visual connections are created by the careful use of glazing in the building. At ground level, extensive windows provide views of the courtyard and the Garden beyond, allowing these internal areas to be read as integral elements of the outdoor landscape.

6. The internal circulation and communal areas focus upon the central court, opening into it at ground level and onto a raised terrace above in order to provide immediate physical connections between the Laboratory and its surroundings.

5

6

TIPS:

1. Limestone
>>> Advantage:
Limestone is readily available and relatively easy to cut into blocks or more elaborate carving. It is also long-lasting and stands up well to exposure.
>>> Disadvantage:
Limestone is reactive to acid solutions, making acid rain a significant problem to the preservation of artifacts made from this stone. Many limestone statues and building surfaces have suffered severe damage due to acid rain. Acid-based cleaning chemicals can also etch limestone, which should only be cleaned with a neutral or mild alkaline-based cleaner.

2. Greek stoa, the monastic cloister
>>> Stoa:
Stoa in ancient Greek architecture are covered walkways or porticos, commonly for public usage. Early stoas were open at the entrance with columns, usually of the Doric order, lining the side of the building; they created a safe, enveloping, protective atmosphere. (Figure a)
>>> Cloister:
A cloister (from Latin claustrum, "enclosure") is a rectangular open space surrounded by covered walks or open galleries, with open arcades on the inner side, running along the walls of buildings and forming a quadrangle or garth. (Figure b, c)

Figure a. The restored Stoa of Attalos in Athens
Figure b. Cloister at Salisbury Cathedral
Figure c. Cloister of Abbaye de Fontenay, in Marmagne, Côte-d'Or, France

South Elevation

Main Lobby Illustrative Section

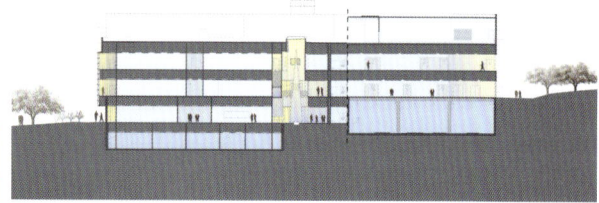

Working Place Transverse Section

Inter-disciplinary learning is facilitated through laboratory layout

| Osborne Centre for Science & Engineering, University of Colorado Colorado Springs

Location:
Colorado Springs, USA
Architect:
NAC Architecture
Construction Area:
14,452 m²
Completion Date:
2009
Photographer:
Frank Ooms

▪ Interdisciplinary laboratory design concept: Interdisciplinary science is one of the fastest growing trends in educational pedagogy and reflects where new scientific discoveries are occurring. Thus, our task was to create formal and informal learning environments for multiple departments that transcend traditional disciplinary boundaries.

▪ The layout of science laboratory:
Inter-disciplinary learning is facilitated through joint projects. Modular lab planning and reconfigurable casework insure adaptability as grant funding and institutional priorities evolve. Informal gathering foster chance encounters between disciplines. These

spaces, located along public circulation, are provided with soft furniture, internet access, and markerboards to facilitate spontaneous meetings. Many of these spaces are placed at the building perimeter to capture daylight while the building's centre is organised around a three-storey skylight lobby with Foucault pendulum.

▪ Overall design concept of a sympathetic campus:
The building was conceived as the means to support place-making and sympathetic campus addition rather than promote an iconic presence.

▪ The project incorporated features to reduce

energy consumption and annual operating costs, such as highly insulated wall/roof systems, a thin film 16.2kw photovoltaic system on the roof, high performance glass, and sunscreens to manage sunlight and heat gain.

The designers also combine energy-saving with the lab's function effectively: the floor plate profile maximises interior daylighting, each elevation was also studied for interior function, campus context, and solar orientation. The result is a building 31% more energy efficient than ASHRAE 90.1 base building.

1. The architecture required deference to existing campus typologies and standards through subtle handling of massing and scale.

2. Fronting the "pedestrian spine", the building nestles into the hillside to provide on-grade entrances to three levels including the main entrance at Plaza.

3. Replacing an existing parking lot with intensive student use provided the opportunity to create vital exterior space that takes advantage of views and solar orientation.

4. By engaging the landform, multiple habitable spaces are provided with overlooks to the campus, city and iconic Pikes Peak.

5. Close-up view of the side surface

6. The spaces, located along public circulation, are provided with soft furniture, internet access, and marker boards to facilitate spontaneous meetings. Many of these spaces are placed at the building perimeter to capture daylight.

7. The public circulation are placed at the building perimeter to capture daylight while the building's centre is organised around a three-storey skylight lobby with Foucault pendulum.

8. Modular lab planning and reconfigurable casework insure adaptability as grant funding and institutional priorities evolve.

9. Combined with view windows into labs and mechanical rooms, the route highlights ongoing research and building system performance.

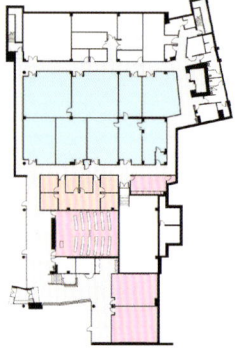

Ground Floor Plan

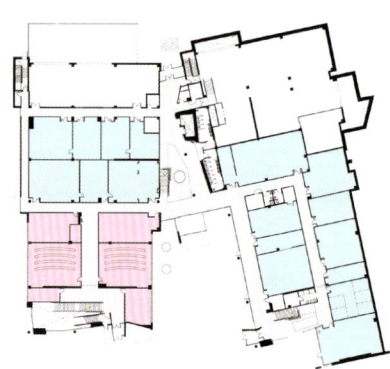

1st Floor Plan

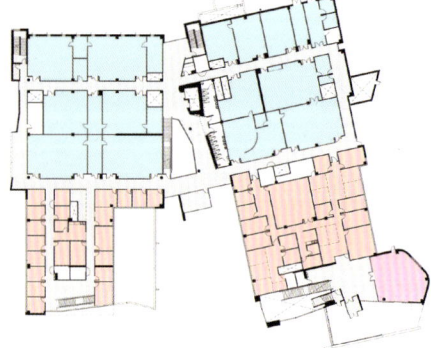

2nd Floor Plan

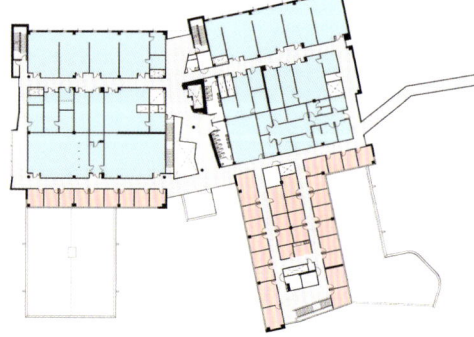

3rd Floor Plan

TIPS:

>>> Awards
The building achieved LEED Gold. Energy design recognition includes the 2009 Institutional Building Award from the Colorado Renewable Energy Society and the CoBiz Magazine Colorado Sustainable Design Award.

>>> Colorado Renewable Energy Society
It is a nonprofit membership organisation whose members work to increase awareness of the economic and environmental benefits of renewable energy and energy efficiency technology and support the sensible adoption of these technologies by Colorado businesses and consumers.
CRES is the Colorado chapter of the American Solar Energy Society.

9

	Office
	Building system
	Circulation
	Lab
	Lab support

Site Section

2

Interactivity, Modularity and Flexibility are underlined by the laboratory space design
| Simon Hall at Indiana University

Location:
Bloomington, Indiana, USA
Architect:
Flad Architects/David Black
Construction Area:
13,045 m²
Completion Date:
2007
Photographer:
Hedrich Blessing Photographers

- Preserving and enhancing the landscape
- Grouping of the labs:
The overall lab design was made with several concepts in mind:
>>Interactivity
>>Modularity
>>Modular utilities
>>Adaptability
>>Flexibility
>>Special areas
In addition to the flexible lab spaces, Simon Hall also provides for specialised core facilities.

By housing these instruments and centres within one central multidisciplinary hub, the university has maximised its investment, allowing multiple researchers access to state-of-the-art technologies. These include:
- A high-field NMR facility containing 800- and 600-MHz NMRs supported by cryo-probes.
- Biosafety level 3 facility.
- An ISO 6 (formerly class 1,000) clean room.
- A high resolution cryo-transmission electron microscope.
- An X-ray crystallography suite.

1. The scale of the new building was kept in harmony with neighbouring historic structures by placing 65,000 square feet of floor space below grade.
2. The building's exterior crafted to respond to their surroundings
3. Subterranean tunnels also connect the building to the three nearby science facilities on the quad, ensuring easy interdisciplinary access to the new labs.
4. Simon Hall is closely aligned with a 1930s art moderne building and incorporates many of that building aesthetic features and design elements.
5. One design element that grew out of consultation with Indiana researchers was the placement of balconies near the labs, for taking in fresh air and short breaks during the long hours required at the lab in the research process.

Office
Building support
Building system
Circulation
Lab
Lab support

Ground Floor Plan

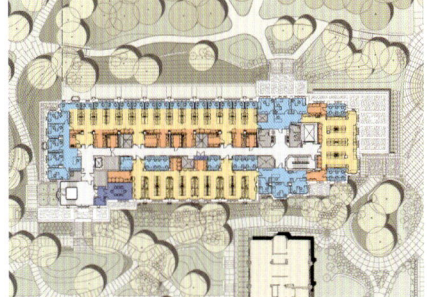

1st Floor Plan

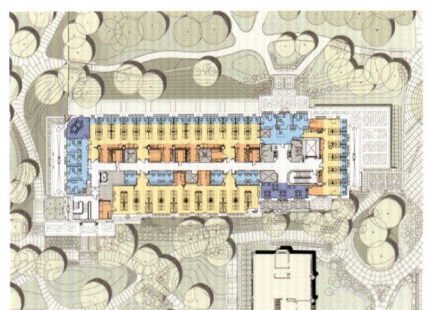

2nd Floor Plan

6. To meet the needs of interdisciplinary research, the modular and flexible labs can be quickly reconfigured to accommodate users from various backgrounds.
7. Flexible layout promotes informal and planned interaction between research colleagues.
8. A variety of specific areas are set to meet different researchers' requirements.
9. Interior stairway

TIPS:

Award name:
R&D Magazine Lab of the Year High Honours
2008
ASLA Wisconsin Honour Award
AIA Wisconsin Merit Award
Outstanding Concrete Achievement Award,
Indiana Ready Mixed Concrete Association
Project of the Year, Masonry Construction

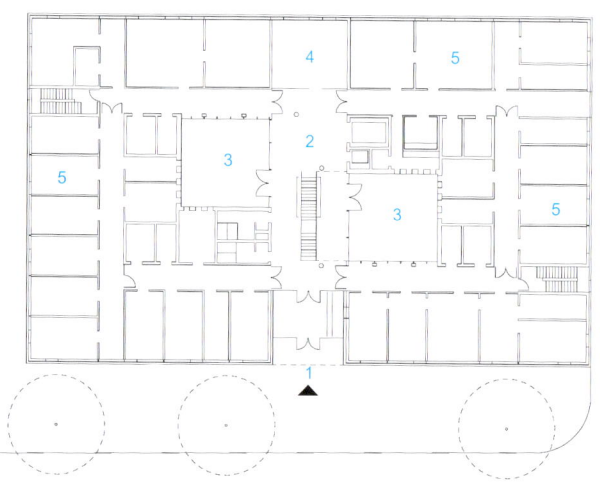

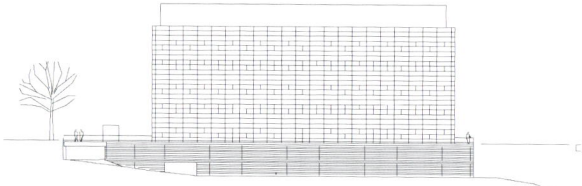

North Elevation

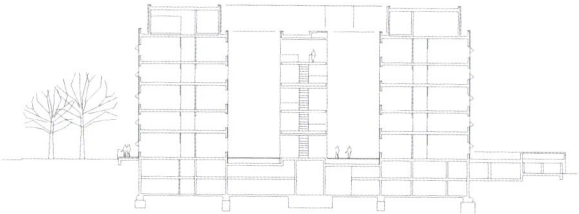

Section

Ground Floor Plan

1. Reception area
2. Foyer
3. Patio
4. Conference room
5. Laboratories

2

The multicoloured fabric outer skin and flexible interior spaces

| BioMedizinZentrum Bochum

Location:
Bochum, Germany
Architect:
hammeskrause architekten
Construction Area:
10,900 m²
Completion Date:
2008
Photographer:
H.G.Esch

■ The challenges in the design process: The designers found themselves confronted with an assignment that is somewhat rare in laboratory construction: to plan a lab building without defined users and hence with no knowledge of their needs.

■ The building's exterior features: The compact structure inserts itself into the homogenous orthogonal fabric of its built urban setting, but with its multicoloured outer skin stands in deliberate contrast to the large stretches of exposed concrete characteristic of the surrounding university buildings.

■ The interior space is flexible: Spatially identical and hence comparable situations were created for all tenants, able to be customised to fit the respective user requirements. The uniform depth enables rooms to be used either as office space or laboratory.

■ Subsidiary spaces such as stairs and courtyard: The five-storey building – sitting atop a two-storey B2 parking garage – alternately layers office and laboratory floors. A lively interplay of interior courtyards, stairways and free airspaces across all storeys creates a central communication space for all the tenants of the new building and provides access to their particular addresses therein.

3

1. The building is set in green surroundings with a wooded slope, in a fringe location on one of the edges of the 1970s campus.

2. Its uniform cellular structure envelops the volume in a skin that unifies the various levels.

3. The articulation of the façade acts as visible expression of the equality of all rental units within.

4. Windows that open outward like ventilation flaps lend the façade a differentiated and individual look.

5. The interior and exterior design convey the exciting tension between a common space shared by several tenants and their diverse individual personalities and activities.

6. Central stairway

4

5

TIPS:

1. The BiomedicineCentre Bochum (BMZ)
The BiomedicineCentre Bochum is located on the campus of the Ruhr University in Bochum and offers 5,200 sqm. of modern laboratory and office space for innovative high-tech companies in medical technology, biomedicine and the healthcare sector.

2. Advantages of the Location
>>>Bochum's strengths in Life Sciences
Bochum has developed into an attractive location for the Life Sciences in the Ruhr region. The city offers companies in the medical technology, biomedicine and biotechnology sectors optimum conditions for founding, starting and watching their business grow. Already there are more than 100 Life Science companies in and around Bochum.

>>>Interdisciplinary Cooperation
Bochum benefits from the interdisciplinary and cross-faculty cooperation of the Ruhr University's medical, natural sciences and engineering departments, which work together successfully on a wide variety of projects. This partnership has led to important special research areas in the Life Sciences being set up at the university.

>>>Central Location
Bochum boasts a central location and proximity to numerous Life Science companies, universities, hospitals, clinics and research facilities in the Ruhr region.

6

1

2

Modular labs, colourful planning
| University of Kansas Medical Centre,
Kansas Life Sciences Innovation Centre

Location:
Kansas City, USA
Architect:
Cannon Design
Construction Area:
19,045 m²
Completion Date:
2007
Photographer:
Michael Spillers Photography

▪ The north, east and west sides of the building exterior are covered with red sandstone precast panels. A hanging canopy and two support columns are clad in zinc, a material that is also used on the standing seam panel roof.

▪ In order to effectively improve the flexible allocation of research groups and the comfort of the lab environment, the designers use efficient modular labs, equipped with lab offices, rest areas and lab suites.

▪ In the design of stairs, to express the colourful working environment, the designers rebuild various organic images in translucent films and put them in the laminated glass. All of the images are selected by the researchers.

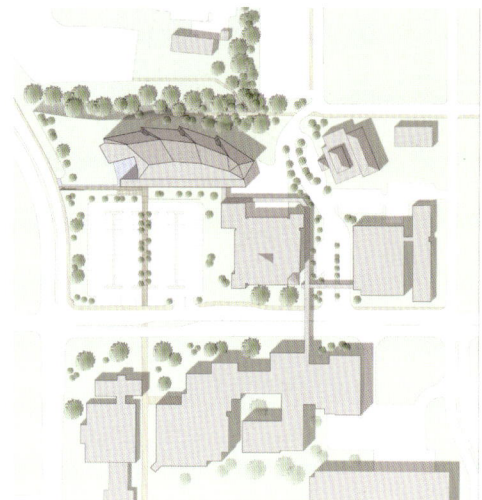

Site Plan

Future Development

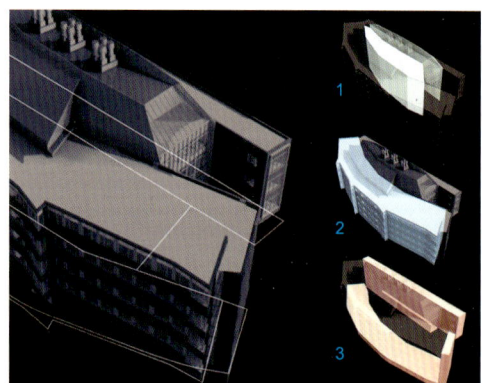

Programmatic Diagrams

1. Core
2. Labs
3. Base

5

6

1. The north, east and west sides of the building exterior are covered with red sandstone precast panels. A hanging canopy and two support columns are clad in zinc, a material that is also used on the standing seam panel roof.

2. The building has a concrete frame structure with a masonry, glass, and precast stone exterior.

3.4. A porous paving grid extending east and west around the north side of the building provides the load-carrying capacity of a paved area with the water-infiltration qualities of natural vegetation.

5. The main entrance of the building faces south. The top 700 square feet of the curtain wall is sloped and fritted with ceramic dots to reduce glare and to prevent ultraviolet rays from penetrating the glass. The roof is 31 feet high and has a maximum slope of 50 degrees.

6. Entrance hall

7. To ensure easy cleaning and decontamination in the state-of-the-art biosafety laboratories, low permeability material was used for the walls and ceilings. In addition, the floors are monolithic (seamless) and slip-resistant, and have coved bases.

8. Efficient, modular laboratories enable flexible assignments of research teams, while "neighbourhoods" of research offices, breakout spaces, and lab suites create dynamic environments for multidisciplinary teaming and focused research centres.

9. The artwork for the stairways was created by reproducing images of various organisms on translucent film and placing them between sheets of glass. The images were selected by the researchers and the result is a colourful example of the work conducted in the building.

10. The ground floor of the building has 4,000 square feet of terrazzo flooring with distinctive geometric patterns in three earth tones.

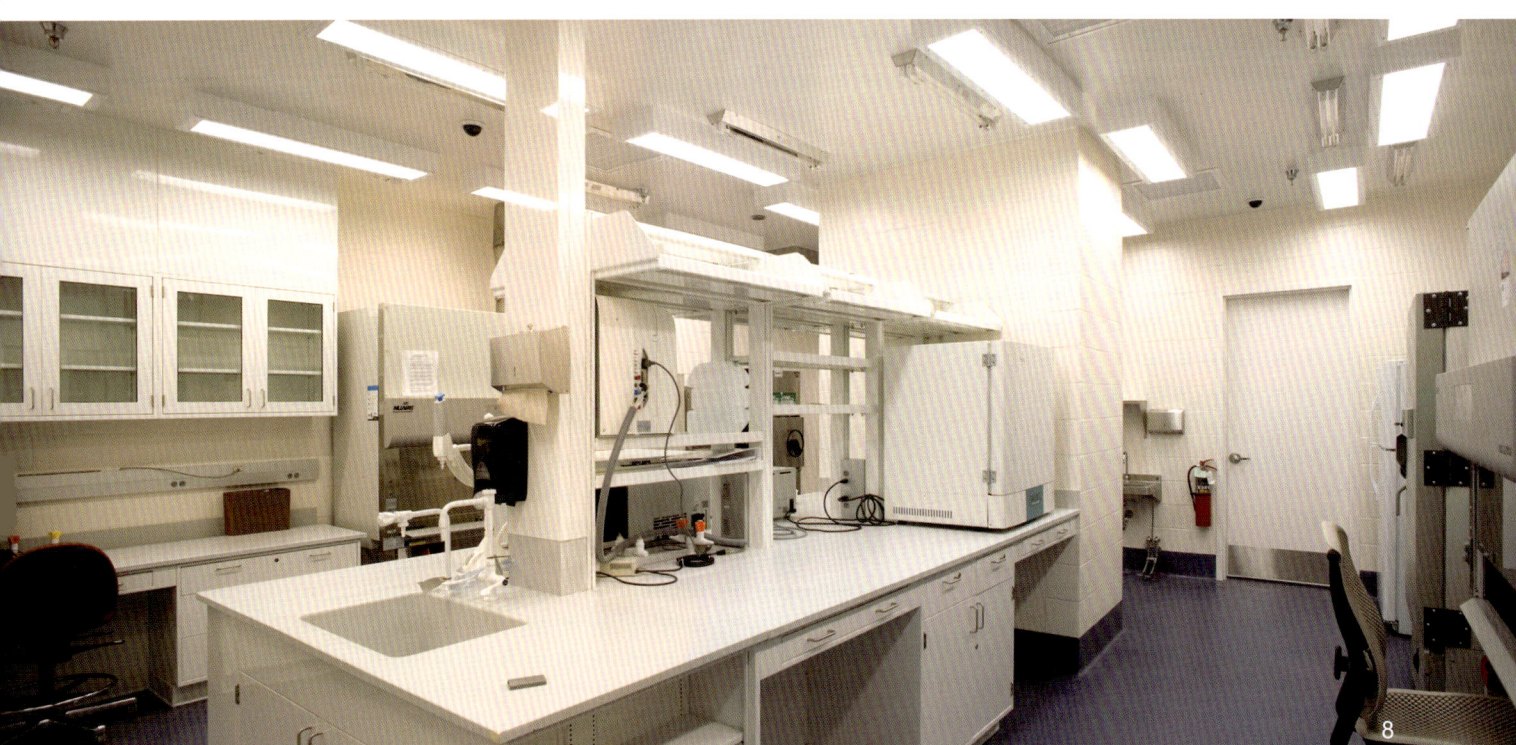

7

8

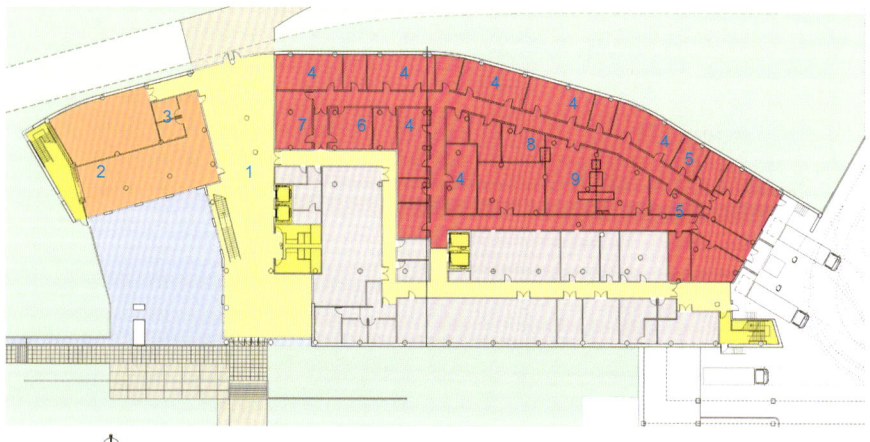

Ground Floor Plan

2nd Floor Plan

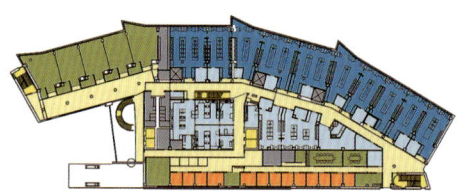

1st Floor Plan

1. Lobby
2. Clinical translational research suit
3. Office
4. Animal hold's room
5. Break room
6. Men's locker
7. Women's locker
8. Sterile store
9. Clean wash

TIPS:

1. Terrazzo flooring
Terrazzo is a composite material poured in place or precast, which is used for floor and wall treatments. It consists of marble, quartz, granite, glass or other suitable chips, sprinkled or unsprinkled, and poured with a binder that is cementitious, chemical or a combination of both. Terrazzo is cured, ground and polished to a smooth surface or otherwise finished to produce a uniformly textured surface.

2. How to make terrazzo floors:
http://www.ehow.com/how_6494634_make-terrazzo-floors.html

3. How to refinish terrazzo floors:
http://www.ehow.com/how_5032532_refinish-terrazzo-floors.html

4. How to make terrazzo floors shine:
http://www.ehow.com/how_6005711_make-terrazzo-floors-shine.html

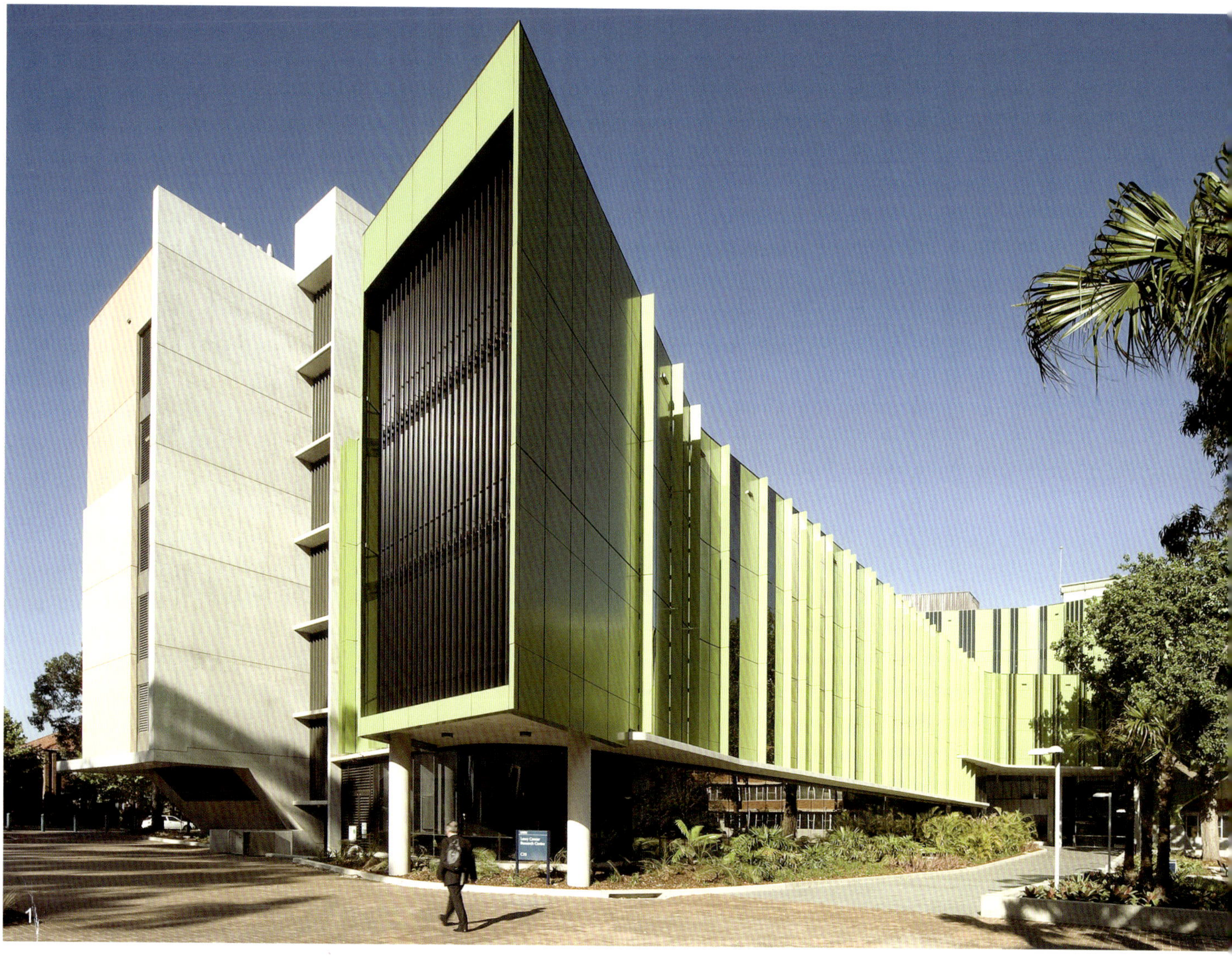
1

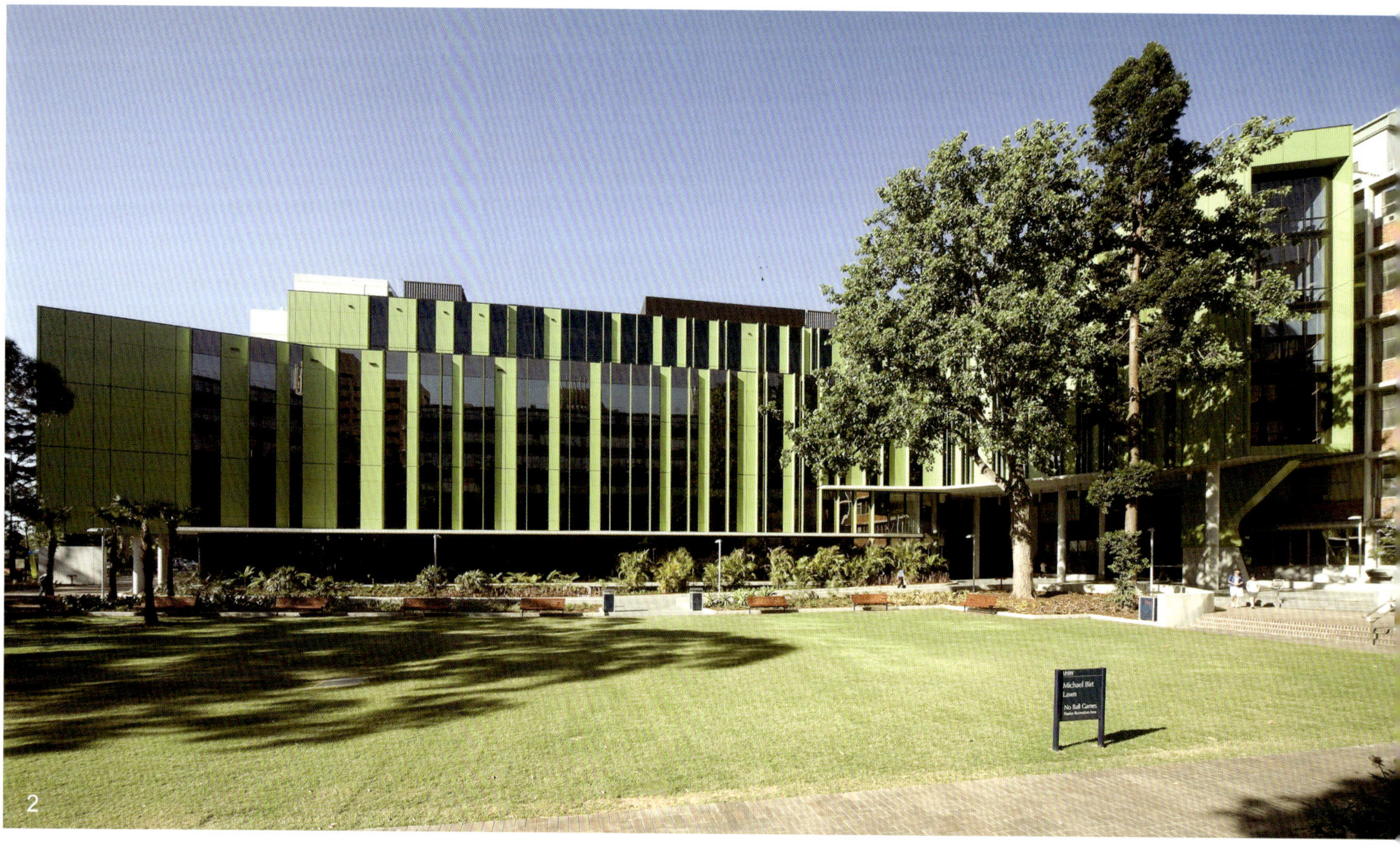

2

3

Laboratory Module design concept can satisfy the requirements of different research organisations

|The Lowy Cancer Research Centre

Location:
Sydney, Australia
Architect:
Lahznimmo Architects, Wilson Architects
Construction Area:
17,000 m²
Completion Date:
2010
Photographer:
Brett Boardman, Anthony Fretwell

▪ The interior design applies "Laboratory Module", which can satisfy the interactive requirements of both small research groups and large research organisations.

▪ The designer takes the atrium as the main social space. The connection of two different functional areas means the closer collaboration between CCIA and FOM.

▪ To reduce its energy use, the designer adopts many existing techniques, including bore water aquifer recharge and cogeneration. This project is assessed as 5 star green architecture by Australia Green Building Committee.

1. The surface is the precast concrete panels and a mix of glazing and green cladding.
2. The building adopts the strategy of expressing "science to the street" and "people to the courtyard".
3. View of the Southern Façade from the Michael Burt Gardens
4. View of the Northwest corner from High Street
5. A bridge links the Lowy Cancer Research Centre to the Wallace Wurth Building on level three.
6. View of the campus gateway entry on High Street
7. Detail view of the campus gateway entry on High Street
8. The main focal and social space in the building is an atrium space, which uses the metaphor of the "science knowledge bank".

6

7

9. The building divides naturally into a more formal but flexible "Laboratory Box" containing labs, support space areas and floor by floor plant.

10. A more fluid "human strand" containing write-up spaces and a variety of areas for break out and collaborative work.

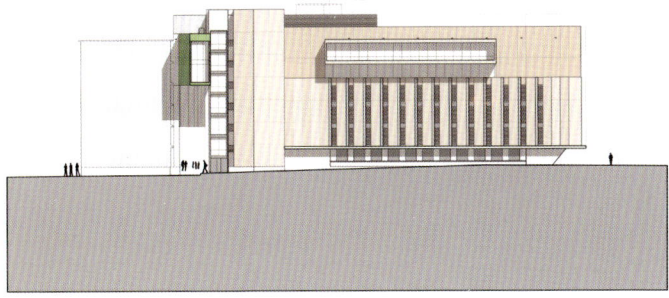

0   10

North Elevation

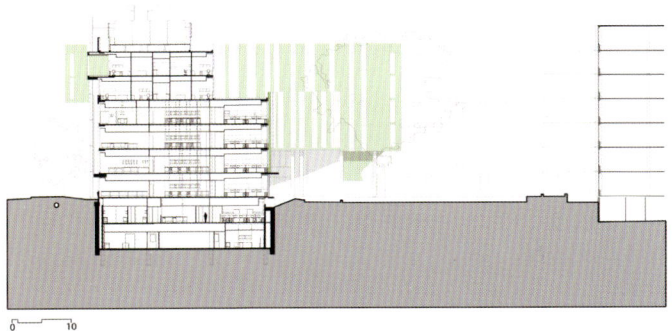

0   10

North-South Section

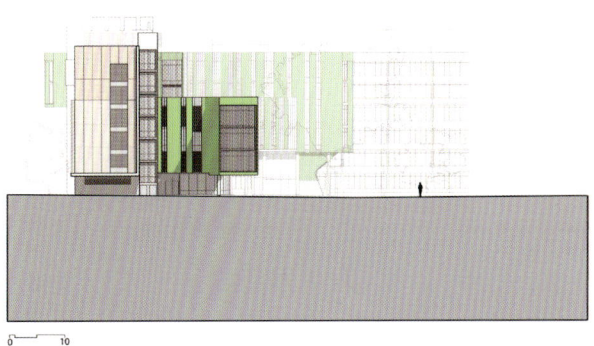

0   10

West Elevation

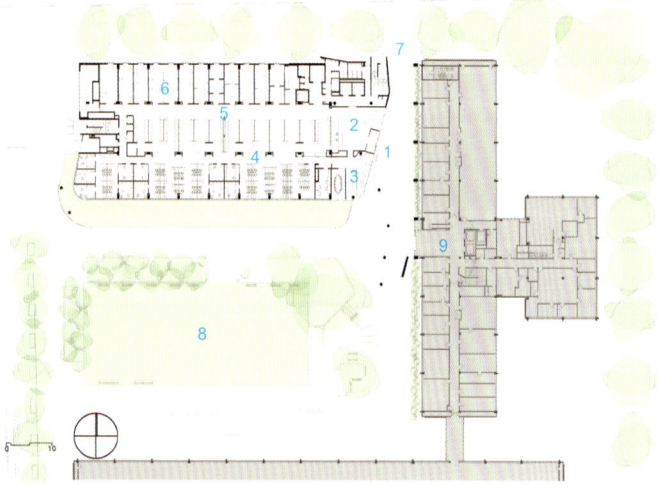

1. Entry
2. Lobby/reception
3. Tea facility
4. Write-up space
5. Laboratory
6. Lab support
7. Campus entry
8. Michael Birt Garden
9. School of Medical Sciences

Site & Ground Floor Plan

TIPS:

1.Aquifer

An aquifer is an underground layer of water-bearing permeable rock or unconsolidated materials (gravel, sand, or silt) from which groundwater can be usefully extracted using a water well. The study of water flow in aquifers and the characterisation of aquifers is called hydrogeology. Related terms include aquitard, which is a bed of low permeability along an aquifer, and aquiclude (or aquifuge), which is a solid, impermeable area underlying or overlying an aquifer. If the impermeable area overlies the aquifer, the pressure could cause it to become a confined aquifer.

>>>Classification:

• Saturated versus unsaturated

• Aquifers versus aquitards

• Confined versus unconfined

• Isotropic versus anisotropic

2. Cogeneration

Cogeneration (also combined heat and power, CHP) is the use of a heat engine or a power station to simultaneously generate both electricity and useful heat.

>>>Types of plants

Topping cycle plants primarily produce electricity from a steam turbine. The exhausted steam is then condensed and the low temperature heat released from this condensation is utilised for e.g. district heating or water desalination.

Bottoming cycle plants produce high temperature heat for industrial processes, then a waste heat recovery boiler feeds an electrical plant. Bottoming cycle plants are only used when the industrial process requires very high temperatures such as furnaces for glass and metal manufacturing, so they are less common.

Large cogeneration systems provide heating water and power for an industrial site or an entire town. Common CHP plant types are:

• Gas turbine CHP plants use the waste heat in the flue gas of gas turbines. The fuel used is typically natural gas.

• Gas engine CHP plants use a reciprocating gas engine which is generally more competitive than a gas turbine up to about 5 MW. The gaseous fuel used is normally natural gas.

• Biofuel engine CHP plants use an adapted reciprocating gas engine or diesel engine, depending upon which biofuel is being used, and are otherwise very similar in design to a gas engine CHP plant. The advantage of using a biofuel is one of reduced hydrocarbon fuel consumption and thus reduced carbon emissions.

• Combined cycle power plants adapted for CHP

• Steam turbine CHP plants use the heating system as the steam condenser for the steam turbine.

• Molten-carbonate fuel cells and solid oxide fuel cells have a hot exhaust, very suitable for heating.

• Nuclear power plants can be fitted with taps after the turbines to provide steam to a heating system. With a heating system temperature of 95°C it is possible to extract about 10 MW heat for every MW electricity lost. With a temperature of 130°C the gain is slightly smaller, about 7 MW for every MW lost.

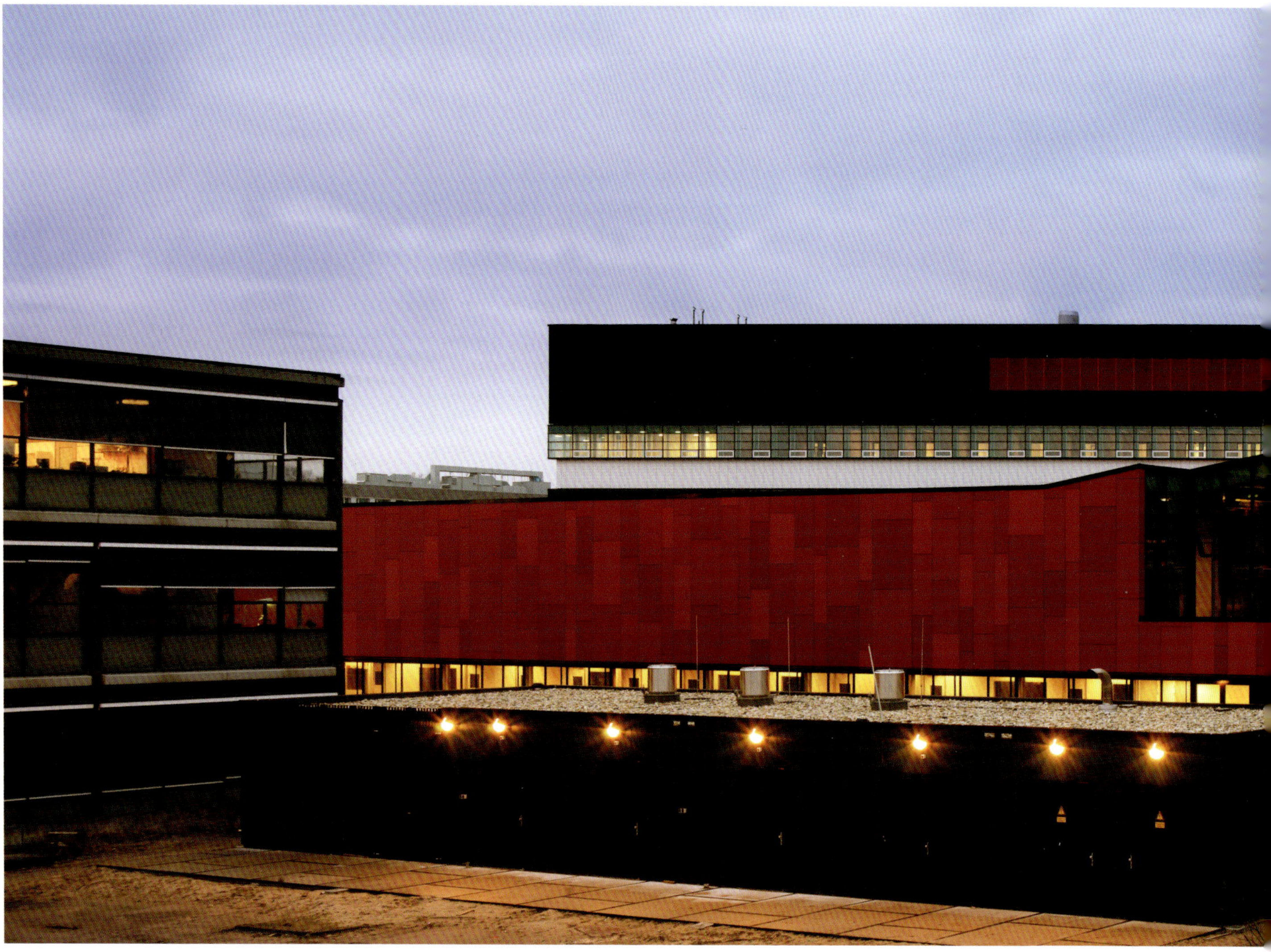

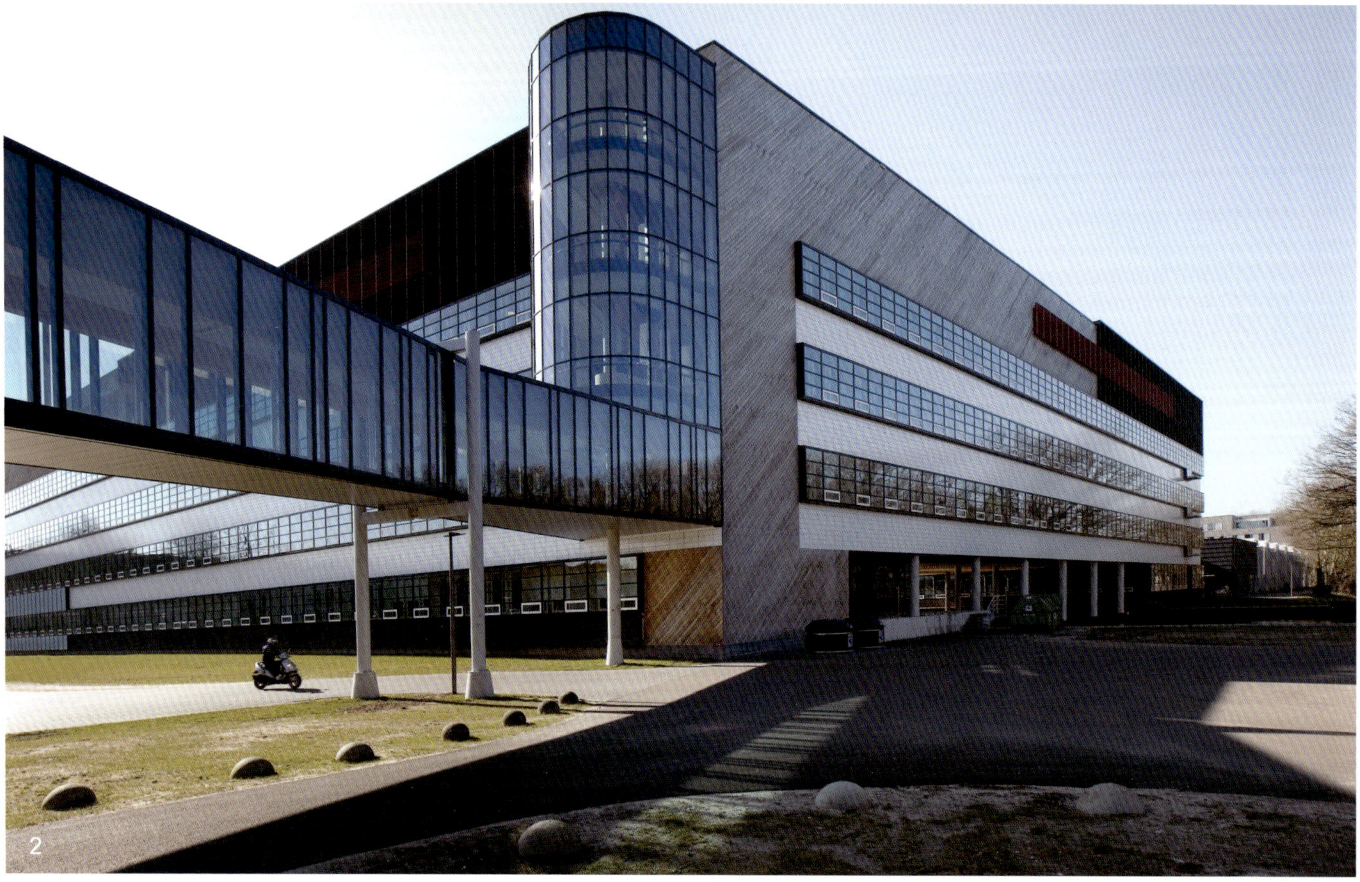

2

With high of flexibility, the interior space is designed without dead end

| Twente University Education and Research Centre with Nanolaboratory

Location:

Enschede, the Netherlands

Architect:

Ector Hoogstad Architecten/Jan Hoogstad (supervisor), Jeroen Huijsinga, Rik van der Velden (architects)

Construction Area:

37,000 m²

Completion Date:

2009

Photographer:

Annet Delfgaauw, Kees Hummel, Marcel van Kerckhoven

▪ Master plan:

The master plan divides up the campus into 2 urbanised parts: one for housing and leisure, the other for education and research. This concentration of building allows large parts of the surrounding park to remain untouched.

▪ The architects were commissioned to design three new buildings in the master plan: the new IT centre , the building for the combined faculties of Science and Technology, Mathematics and Computer Sciences (Carré) and the Institute for Nanotechnology (Nanolab). The latter two together form the heart of the Education and Research Centre.

▪ To provide a flexible interior space, the designers used continuous orthogonal routes to ensure that the corridors have no dead ends.

3

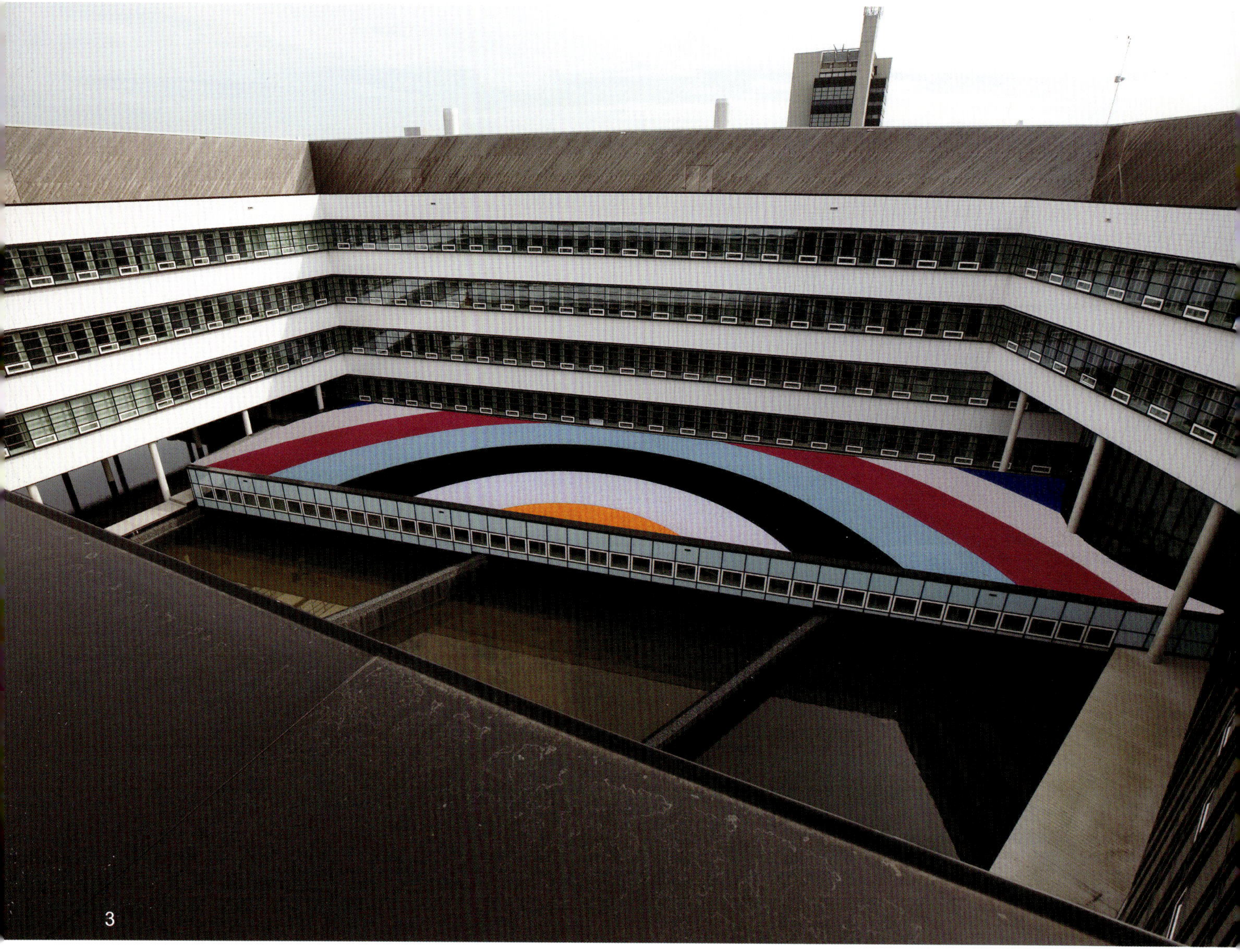

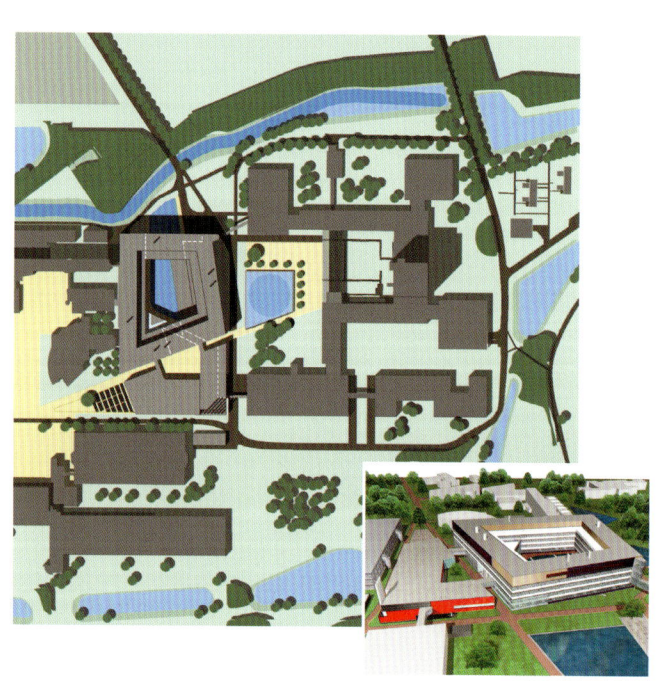

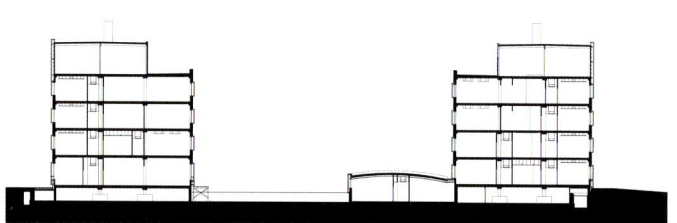

Carré Section

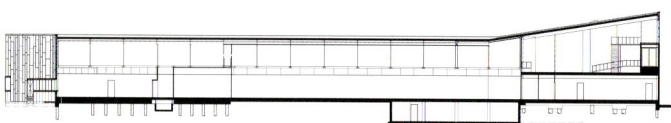

Nanolab Section

Site Plan

1.2. The Carré together with the new Nanolab form an ensemble that connects all existing buildings that make up the Education and Research centre.

3. The design has a singularly deep nave of 21.6 metres. This gives the building optimum flexibility, while the continuous rectangular routing facilitates smooth passage without any dead ends.

4. Carré is a five-storey building enclosing a central courtyard, which houses all office and laboratory facilities of the 2 faculties.

5. A wide transparent corridor, crossing the courtyard at the level of the ground floor provides access to the adjacent Nanolab.

6. Entrance hall

7. Traffic stairs in the Carré

8. Hallway in the Carré

9. The Nanolab accommodates the most valuable laboratories the university has to offer.

6

7

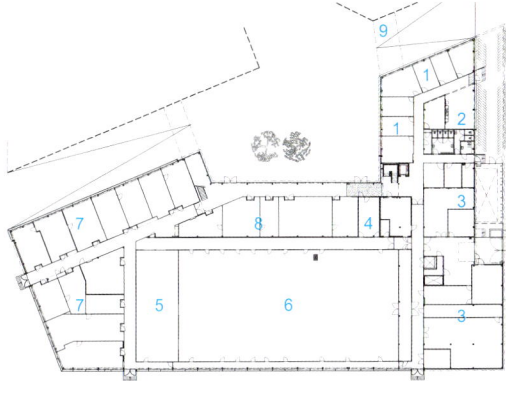

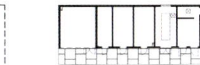

Ground Floor Plan of Nanolab

1. Office
2. Social meeting point
3. Technique
4. Kleedsluis
5. VC-G clean rooms
6. VC-D clean room
7. VC-G labs
8. VC-D labs
9. Bridge

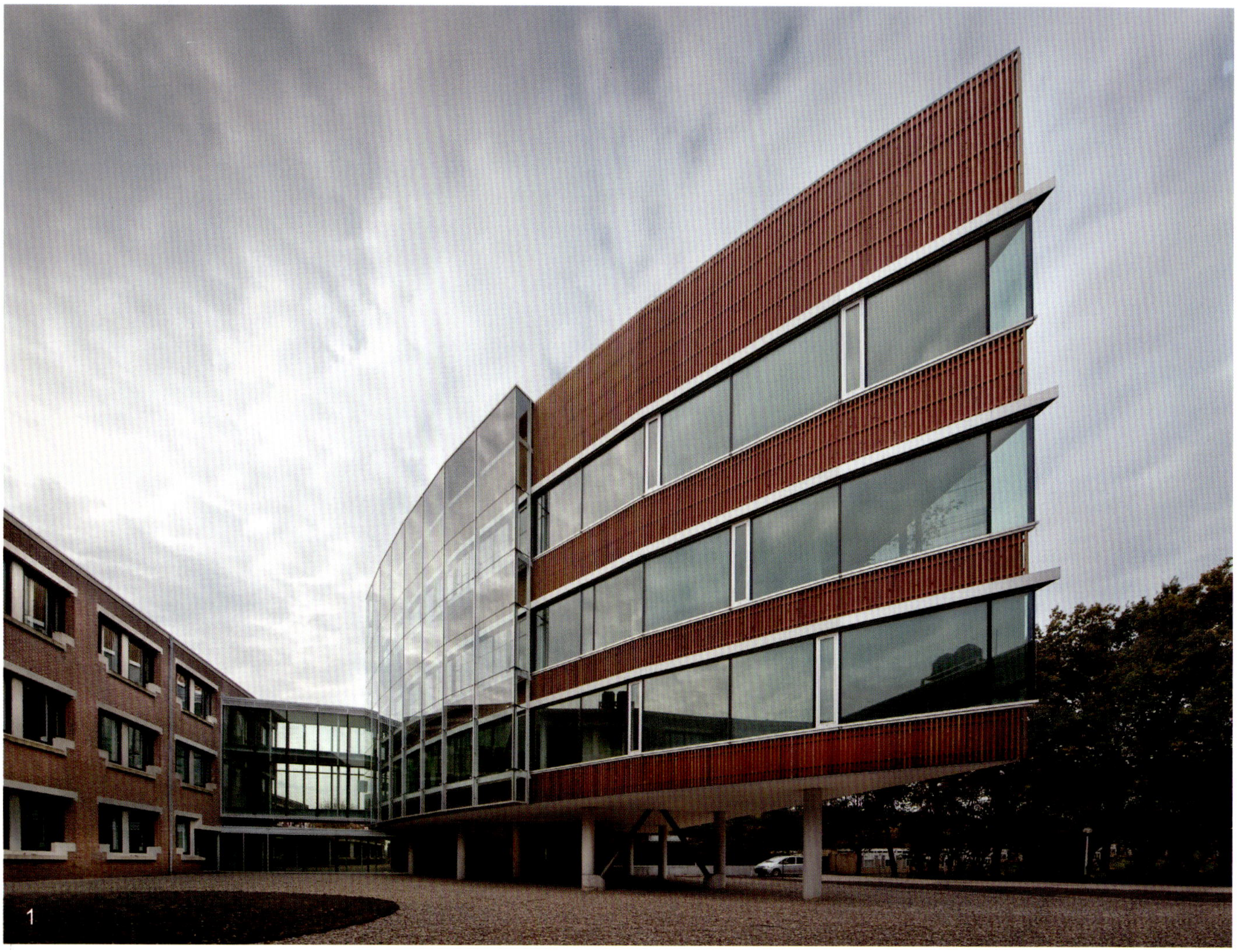

1

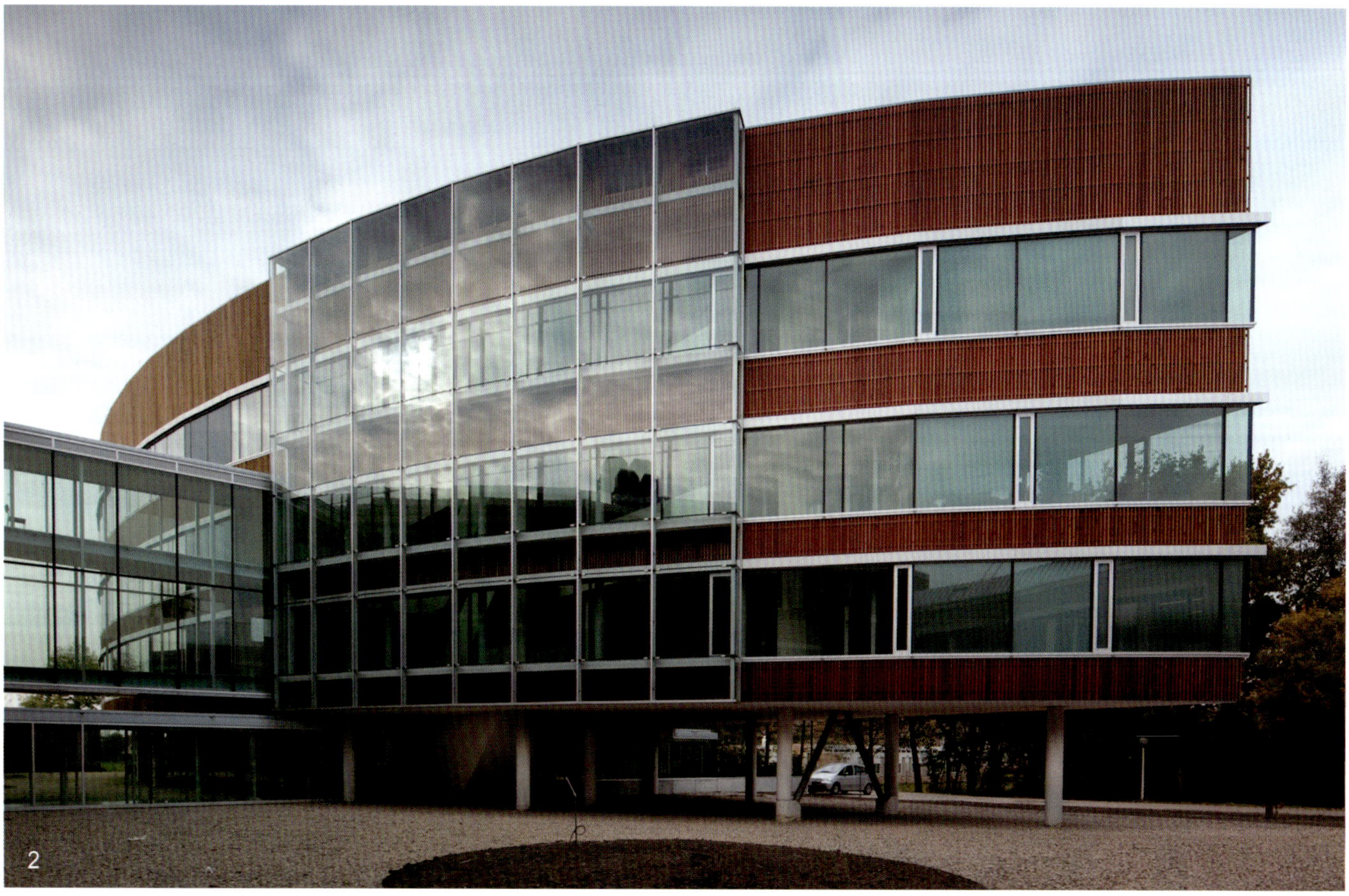

2

The uniform "half-moon" shape and the red exterior impart a vivacious façade

| RIKILT/VWA Wageningen

Location:
Wageningen, the Netherlands
Architect:
Broekbakema/Erik van Eck, ir. Oscar van Strijp, Kees van Zwol
Construction Area:
Renovation 6,195 m², New building 4,430 m²
Completion Date:
2009
Photographer:
Menno Emmink

▪ The whole building design philosophy: It was decided to make the new construction stand out both in terms of form and personality.

▪ Shape and façade: The uniform "half-moon" shape of the new construction makes a clear gesture to the campus, while the curved façade lends a certain largesse and view on the existing building.

▪ The vivacious and unpretentious façade: With its divergent storey heights to house the offices (3,600mm gross) and the labs (4,200mm gross) and large windows starting from the floor, the superstructure imparts a vivacious yet unpretentious façade. The timber beams placed at the regular distance on the red high pressure laminate board create a technical yet natural and warm façade.

▪ Greenspace and landscape: Outside, the greenspace runs under the new construction to meet the existing lab building. Using landscaping elements, the fragmented elevations on the terrain are blended into a whole.

4

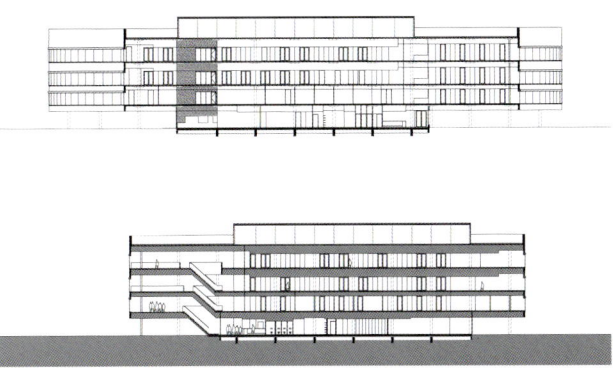

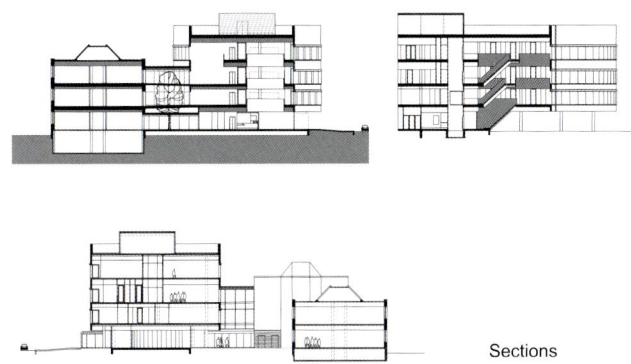

Sections

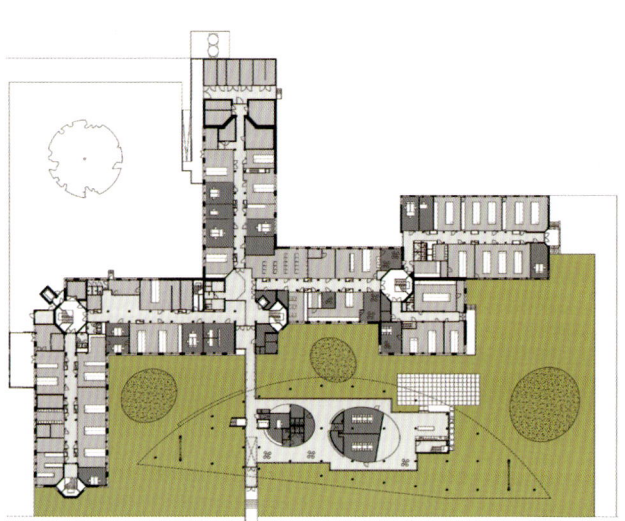

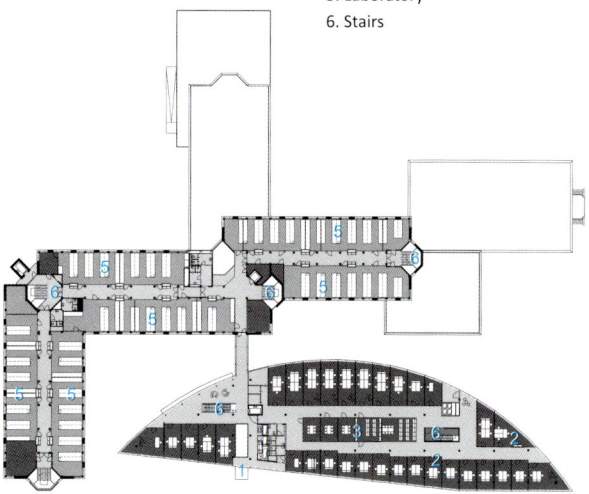

1. Entrance
2. Office area
3. Conference rooms
4. Café
5. Laboratory
6. Stairs

Ground Floor Plan

1st Floor Plan

1. The curved façade lends a certain largesse and view on the existing building. The resulting effect is a pleasant space between old and new.
2. By standing the building on feet and giving the substructure a transparent form containing the most representative and public functions, a certain liveliness is created as well as an open view on the existing building at street level.
3. Outside, the green space runs under the new construction to meet the existing lab building. Using landscaping elements, the fragmented elevations on the terrain are blended into a whole.
4. The timber beams placed at the regular distance on the red high pressure laminate board create a technical yet natural and warm façade.
5. The new construction is, as it were, nestled in the "armpit" of the existing laboratory and is linked to it by a number of bridges.
6. Connection bridge
7. Reception and hall
8. The staircase in the main hall
9. Laboratory

8

TIPS:

Education & research of Wageningen
In 1918 the town acquired its first institution of higher education, Netherlands Agricultural College which later became Wageningen University.

Today, Wageningen is also the central city in Food Valley, the Dutch food & nutrition cluster concentrated around WUR and comprising many institutes, companies and state-of-the-art facilities in the food & nutrition field. Food Valley is regarded as the largest food & nutrition Research & Development cluster in the world. One such firm, Keygene, a biotechnology company in Wageningen developed AFLP in the early 1990s and collaborated with Beijing Genomics Institute to sequence the entire genome of Brassica napus.

9

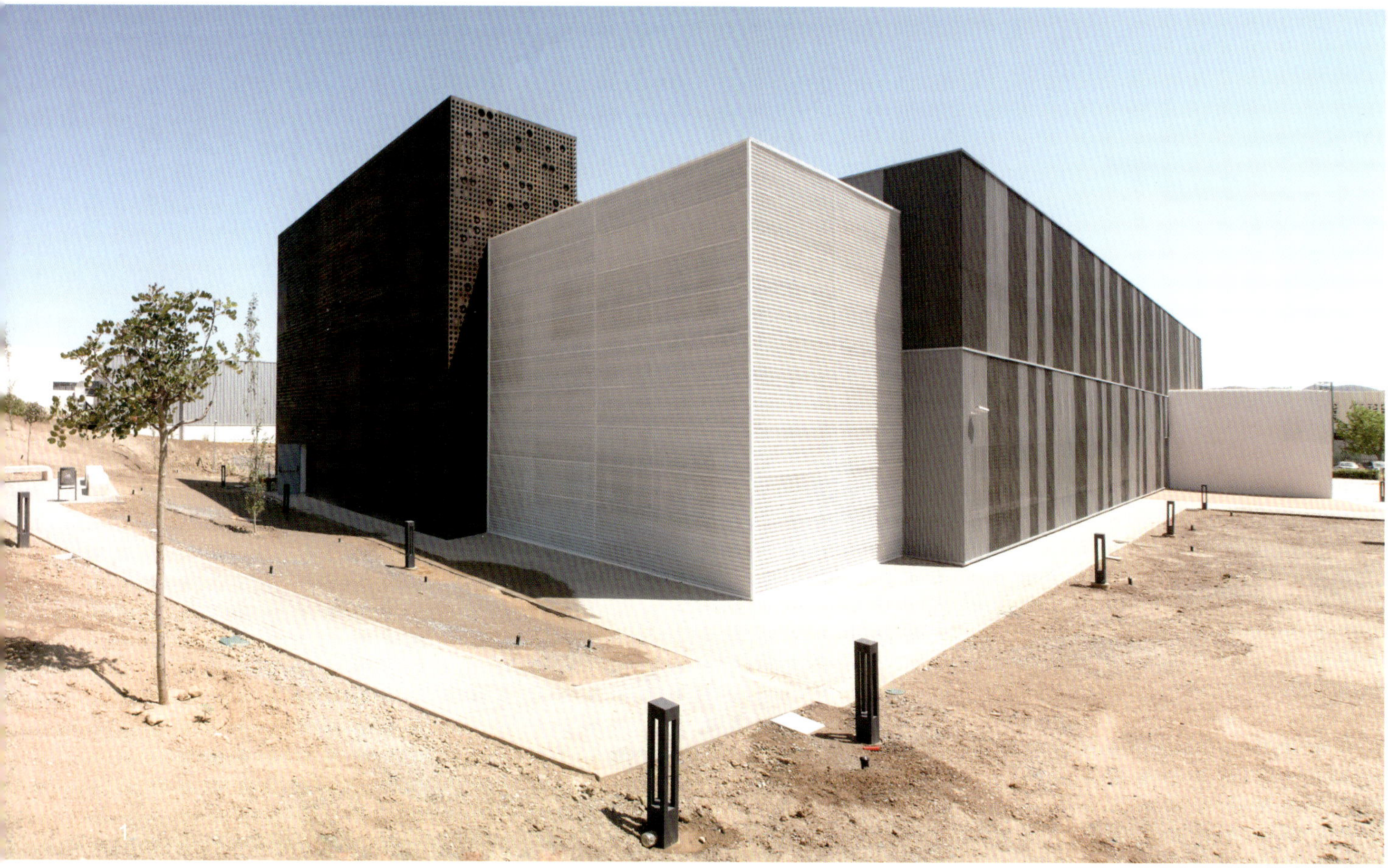

1

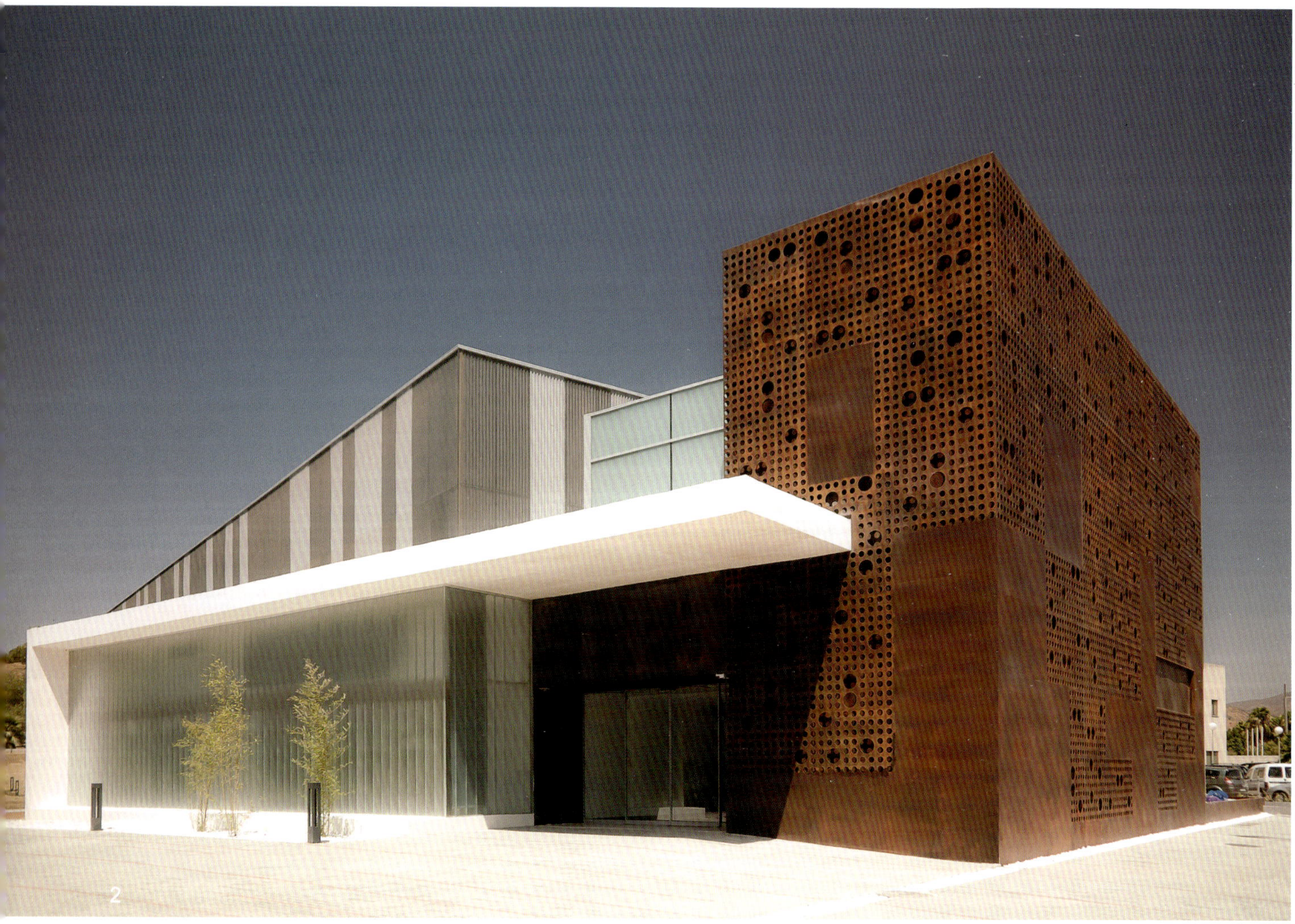

2

The design of patio-skylights improves the interior lighting

| The Andalusian Centre for Nanomedicine and Biotechnology

Location:
Málage, Spain
Architect:
Planho Consultores
Construction Area:
6,500 m²
Completion Date:
2010
Photographer:
Alejandro González

■ The use of perforated steel panels ensures the solid image of the building.
■ Overall arrangement of the inteior space: The designer arranges the labs in the outer area, and the support spaces and reseacher offices in the centre. The labs are seprated by simple glass panels, providing compactedness and flexibility for the interior.

Based on the compact and flexible structure, the deisgn of patio-skylights improves the interior lighting effect significantly.

Double corridor model: It is the ideal scheme for functional structures that require changes without affecting the architectural elements.

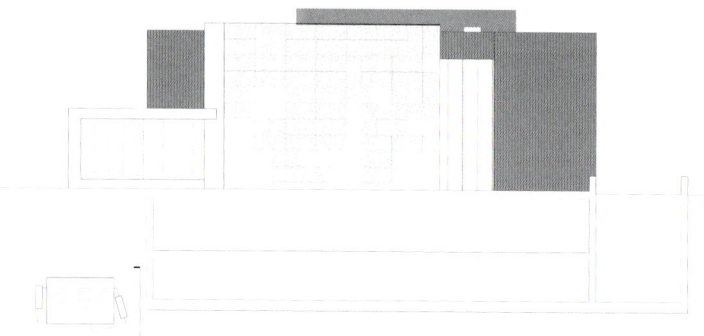

East Elevation

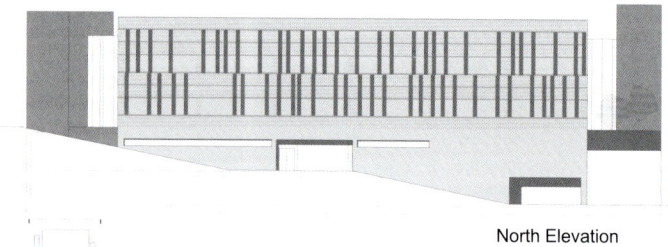

North Elevation

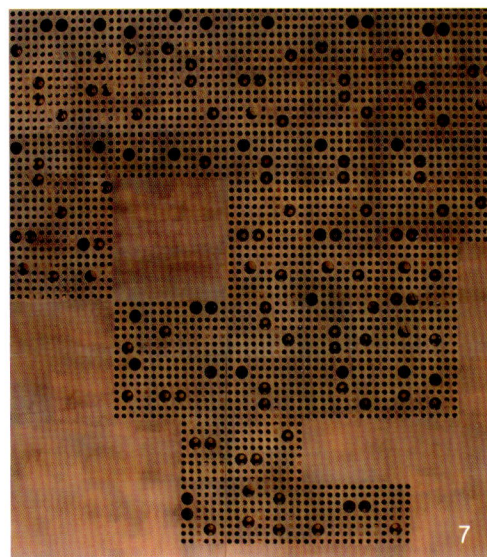

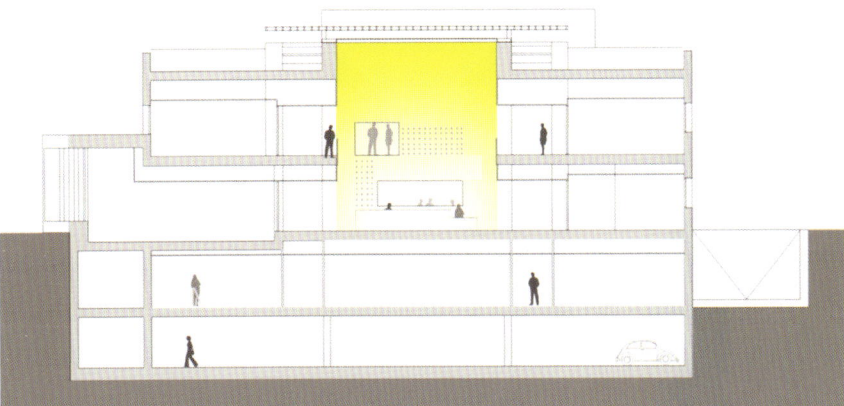

Section

1. The Bionand building is a basically horizontal structure, with a triple centreline scheme.

2. The building locates the laboratories in the façades and the support spaces and investigators in the central zone, lit up by the large cavities in the patio-skylights.

3. The building consists of 4 floors, and the research facilities are developed on the top two floors, leaving the semi-basement for support and service zones and the basement for parking.

4. Main entrance

5. Façade of the entrance

6. With regard to the materials, the building combines Corten steel in core areas with perforated aluminium sheets.

7.8. Details of aluminium sheet

9.10. The patio-skylights plays a central role in this project. It introduces necessary space in an intentionally dense structure, brings natural lighting to every single corner of the building, and is an essential aspect in the perception and comfort of the general areas.

11.12. The matrix scheme with double corridor allows optimum use of resources as well as versatility in the use of the research and support facilities.

13. The laboratories has a completely flexible organisation, continuing on from each other, simply separated by a glass panel.

14. The location of the facilities plays an important role in the versatility of this configuration, which guarantees that the building serves as a flexible container for any kind of work that one wishes to implement.

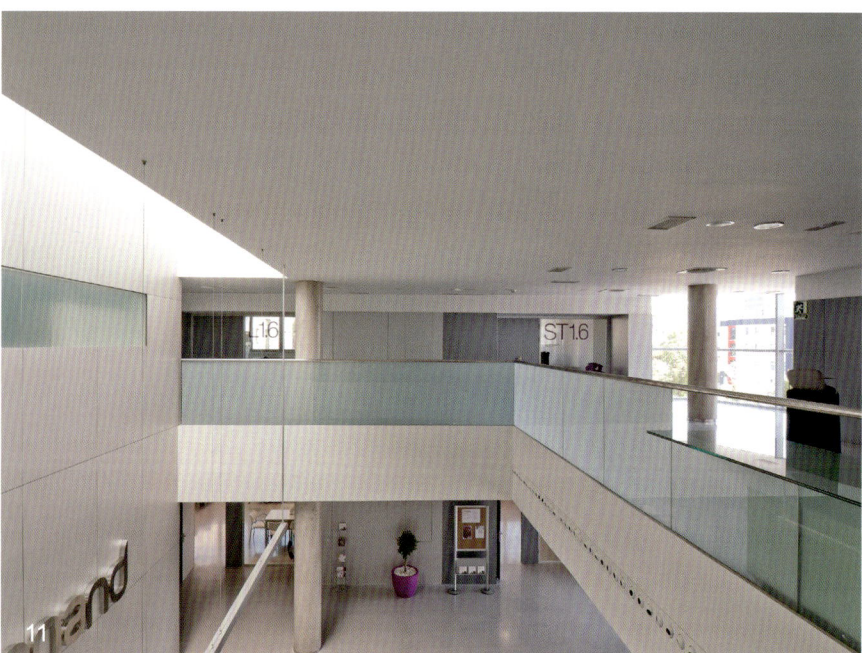

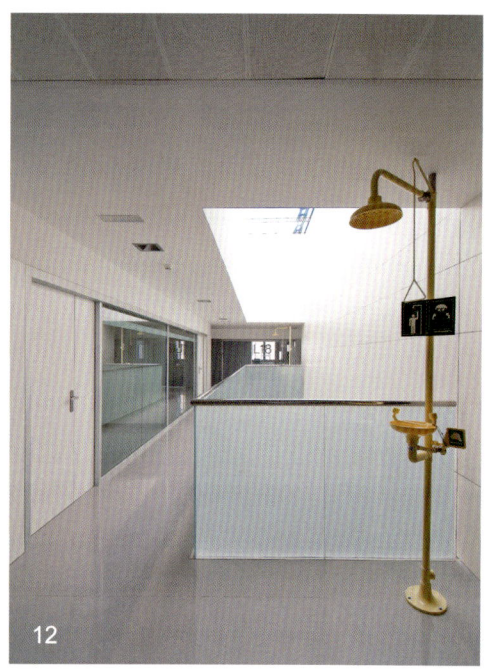

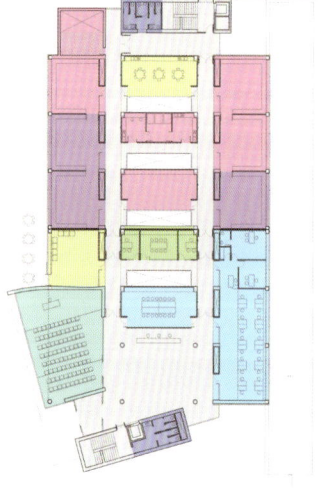

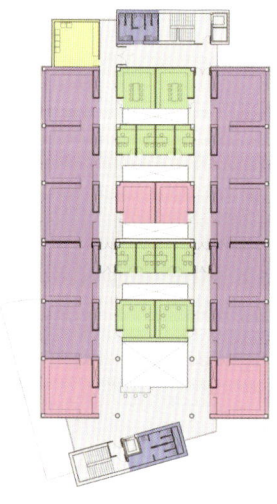

■	Laboratories
■	Resonance
■	Staff
■	Teaching
■	Supports
	Parking

Basement Plan

Ground Floor Plan

1st Floor Plan

13

TIPS:

Corten steel

Weathering steel, best-known under the trademark COR-TEN steel and sometimes written without the hyphen as "Corten steel", is a group of steel alloys which were developed to eliminate the need for painting, and form a stable rust-like appearance if exposed to the weather for several years.

"Weathering" means that due to their chemical compositions, these steels exhibit increased resistance to atmospheric corrosion compared to other steels. This is because the steel forms a protective layer on its surface under the influence of the weather.

The corrosion-retarding effect of the protective layer is produced by the particular distribution and concentration of alloying elements in it. The layer protecting the surface develops and regenerates continuously when subjected to the influence of the weather. In other words, the steel is allowed to rust in order to form the "protective" coating.

Challenges: Using weathering steel in construction presents several challenges. Ensuring that weld-points weather at the same rate as the other materials may require special welding techniques or material. Weathering steel is not rustproof in itself. If water is allowed to accumulate in pockets, those areas will experience higher corrosion rates, so provision for drainage must be made. Weathering steel is sensitive to salt-laden air environments. In such environments, it is possible that the protective patina may not stabilise but instead continue to corrode.

14

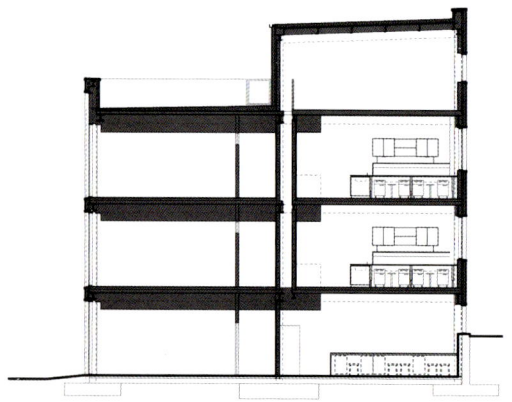

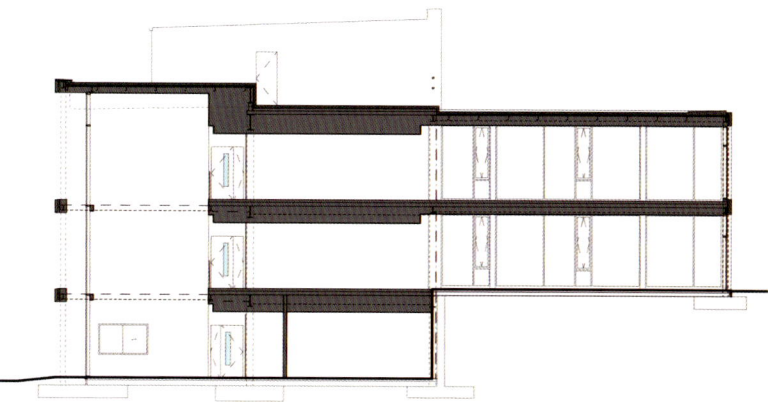

Cross Sections

The floor-to-ceiling windows take full advantage of the natural light and the interior is designed to promote inter-disciplinary communication

| CH Waddington Building, University of Edinburgh

Location:
Edinburgh, Scotland, UK
Architect:
HOLMES MILLER
Construction Area:
1,600 m²
Completion Date:
2009
Photographer:
Andrew Lee

▪ Holmes Miller's design takes full advantage of the attractive location of the site which benefits from an intimate garden setting on ground level and impressive views to the south-east of the city and beyond on the upper levels.
▪ The three-storey building provides a pleasant working environment with maximum use of natural light through floor-to-ceiling windows.
▪ Simplicity is elegant. The simple gridded fenestration results in a very elegant overall elevational composition.

▪ Characteristics of interior spatial organisation: Flexibility and adaptability.
"Play Spaces" with seating and blackboards are provided to encourage informal interaction between researchers.
A key element of the brief was the interface with the adjacent Daniel Rutherford Building, nurturing and supporting communication between the two facilities whilst allowing staff to meet and chat informally within the circulation spaces.

1. Main entrance
2. Locker
3. Administration
4. Copy
5. Office
6. Meeting

7. Specialist laboratory
8. Seminar
9. Ancillary
10. Break space
11. Cold room

Ground Floor Plan

4

1. Beyond the functional areas, the client's vision, clearly stated at the outset, was for a naturally lit environment which encourages both formal and informal inter-disciplinary contact.

2. The building exudes a quite confidence through appropriate scale and massing and a simple grid fenestration which sits comfortably alongside the existing Daniel Rutherford Building.

3. The three-storey building provides a pleasant working environment with maximum use of natural light through floor-to-ceiling windows and social areas at ground floor level for enjoyment and use of the gardens.

4. The grid fenestration details

5. A degree of flexibility and adaptability, in terms of internal partitioning, has been incorporated into the building's design to allow for the changes in future requirements and the evolving nature of the users' research programmes.

6. Within the circulation zones "Play Spaces" with seating and blackboards are provided to encourage informal interaction between researchers.

7. The classroom is full of natural light through floor-to-ceiling windows.

7

TIPS:

Client Quote:
"Key components of the brief for our new home were the delivery of integrated office space, embedded social areas and shared laboratory provision. I can confirm that all of these elements were delivered with considerable success by the Holmes Miller Partnership design. Our offices allow interaction between previously disparate staff; several new working relationships have arisen from these new office arrangements. The embedded social space has proved to be very effective; careful design has allowed the space to function well... Finally, the shared laboratory areas have pooled staff, equipment and expertise into a cutting-edge laboratory."

Dr Elisabeth Elliot, Centre Manager (CSBE)

A new model of laboratory design is emerging, one that creates lab environments that are responsive to present needs and capable of accommodating future demands. Several key needs are driving the development of this model:

Creating a "Social" Research Space Based on Collaborative Research

Modern science is an intensely social activity. The most productive and successful scientists are intimately familiar with both the substance and style of each other's work. They display an astonishing capacity to adopt new research approaches and tools as quickly as they become available. Thus, science functions best when it is supported by architecture that facilitates both structured and informal interaction, flexible use of space, and sharing of resources.

1. Communication Spaces

A critical consideration in designing such an environment is to establish places – break rooms, meeting rooms, atrium spaces – where people can congregate outside their labs to talk with one another. Even stairways, fire stairs, or stairs off an atrium with built-in window seats can provide opportunities for people to meet and exchange ideas. Designers must look for opportunities for such uses in public spaces, making optimal use of every square foot of the building. (See Figure 1)

2. The Laboratory Based on Team (See Figure 2)

Collaborative research requires teams of scientists with varying expertise to form interdisciplinary research units. As networks connect people and organisations, sharing data within a team and with other research teams becomes less complicated. So, designers are organising space in new ways. Laboratory designers can support collaborative research by:

• Creating flexible engineering systems and casework that encourage research teams to alter their spaces to meet their needs.

• Designing offices and write-up areas as places where people can work in teams.

• Creating "research centres" that are team-based.

• Creating all the space necessary for research team members to operate properly near each other.

• Minimising or eliminating spaces that are identified with a particular

Figure 3. The laboratory space is not only designed with team-based open concepts but also provides convinces for independent research activities.

department.
- Establishing clearly defined circulation patterns.
- Provide interior glazing to allow people to see one another.

The Balance between "Closed" Space and "Open" Space

An increasing number of research institutions are creating "open" labs to support team-based work. The open lab concept is significantly different from that of the "closed" lab of the past, which was based on accommodating the individual principle investigator. In open labs, researchers share not only the space itself but also equipment, bench space, and support staff. The open lab format facilitates communication between scientists and makes the lab more easily adaptable for future needs. A wide variety of labs – from wet biology and chemistry labs, to engineering labs, to dry computer science facilities – are now being designed as open labs.

Most laboratory facilities built or designed since the mid-1990s in the U.S. possess some type of open lab.

There can be two or more open labs on a floor, encouraging multiple teams to focus on separate research projects. The architectural and engineering systems should be designed to affordably accommodate

Closed labs are still needed for specific kinds of research or for certain equipment. Nuclear magnetic resonance (NMR) equipment, electron microscopes, tissue culture labs, darkrooms, and glass washing are examples of equipment and activities that must be housed in separate, dedicated spaces. (See Figure 3)

Moreover, some researchers find it difficult or unacceptable to work in a lab that is open to everyone. They may need some dedicated space for specific research in an individual closed lab. In some cases, individual closed labs can directly access a larger, shared open lab. When a researcher requires a separate space, an individual closed lab can meet his or her needs; when it is necessary and beneficial to work as a team, the main open lab is used. Equipment and bench space can be shared in the large open lab, thereby helping to reduce the cost of research. This concept can be taken further to create a lab module that allows glass walls to be located almost anywhere. The glass walls allow people to see each other, while also having their individual spaces.

Creating a Humane and Flexible Space

Maximising flexibility has always been a key concern in designing or renovating a laboratory building. Flexibility can mean several things, including the ability to expand easily, to readily accommodate reconfigurations and other changes, and to permit a variety of uses.

In this electronic/information age, work teams form and reform to meet organisational needs, technological innovations, and changing business relationships. Buildings and interior spaces need to be flexible to anticipate and support this changing nature of work. Within the past few years, designers have sought to create a new generation of "flexible" buildings and workplace environments within buildings that have infrastructures and structures that fully support change while sustaining new technologies, and multi-capable individuals and teams.

The changing nature of work means greater mobility for workers, a multiplicity of workspaces within and external to buildings, greater use of geographically dispersed groups, increased dependence on social networks – and greater pressure to provide for all of these needs and behaviors in a leaner and more agile way. Workplaces have responded with many new options, including more teaming and informal interaction spaces, more supports for virtual individual and group work, more attention to integrating learning into everyday work experience, greater flexibility in work locations, and more focus on fitting the workplace to the work rather than vice versa. Many workplaces are also incorporating spaces that encourage relaxed engagement with colleagues to reduce stress and promote a sense of community. (See Figure 4)

Recommendations as followings:
• Consider wireless technology and mobile phones to enable workers to move effortlessly among spaces as their needs change.
• Consider the use of dedicated project rooms for some types of group work.
• Provide multiple places to meet and greet.
• Consider providing informal workspaces in cafeterias.
• When designing cafés and coffee nooks, locate them centrally along well traveled pathways to encourage use and interaction. (See Figure 5)
• Design the circulation system with informal communication opportunities in mind. (See Figure 6)
• Consider sharing meeting spaces among private offices.
• Support stress reduction and relaxation. (See Figure 7)

4

Figure 4. The research space is created with more flexibility to allow for team-based research and informal interaction.
Figure 5. The resting area is located along well traveled pathway to encourage interaction between the employees.
Figure 6. Informal communication space in the corridor
Figure 7. Design the corridor with some resting areas

The Spatial Coordination between Lab Facilities and Computer System

One important change that has occurred in the design of research facilities is that furniture must be designed with computer use in mind. For example, furniture must accommodate the cabling necessary for PCs or laptop computers. Tables should be modular so that they can be added to or rearranged consistent with the fixed casework and the lab equipment to meet criteria for the space. Ports and outlets should be located to accommodate multiple furniture layouts. Write-up stations should to be at least four ft wide to allow for knee space and hardware under the countertop.

Workstations should be 48 in. wide and 30 in. deep, at a minimum. If a computer will be shared, the workstation should, at a minimum, be 72 in. wide and 30 in. deep. In wet labs, computer keyboards must be placed away from spill areas, ideally in separate write-up areas. Laptop computers should be considered for their compact size, mobility, and ease of storage. Electrical outlets must be accessible for plugging in adapters. And, as was mentioned in an earlier section, designers should consider stacking hardware vertically on mobile carts. Laptops with voice-activated microphones are being developed for use in fume hoods, where use of standard laptops can create safety hazards (or where laptops might be damaged by chemical spills). (See Figure 8)

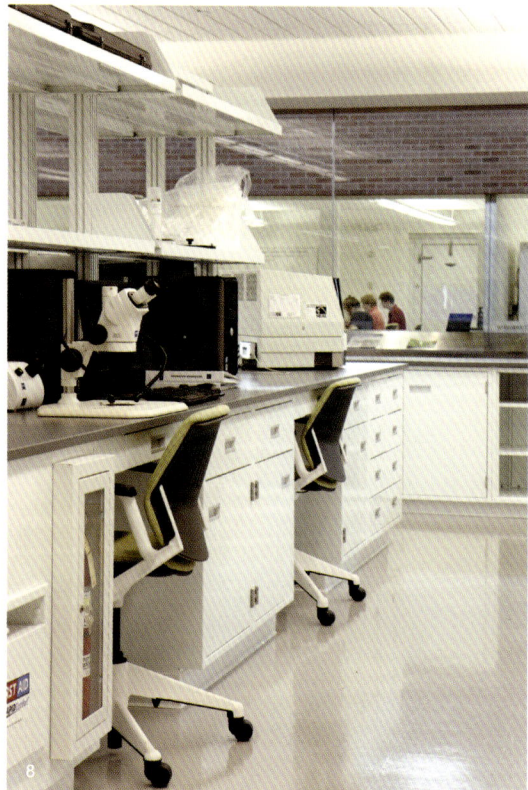

Figure 8. Laboratory facilities and computer system are reasonable arranged.

Three key developments in computer furniture should be emphasised:
• Specialised equipment enclosures.
• Computer hardware enclosures. Hardware enclosures that are fully ventilated and secure are available. Security for computers in a lab is a management and design issue, and designers should consider mobile cabinets with adjustable shelving that can be locked.
• Monitor arms, server platforms, and keyboard drawer solutions. Monitor arms are capable of holding up to 100 lbs. and can support computer monitors of up to 21 in. Mobile server platforms are designed with adjustable shelving to allow stacking of computer hardware. Keyboard platforms can be adjusted vertically and can be mounted under the work surface.

Throughout the research industry today, one constantly hears the phrases "virtual labs" and "virtual reality". In the most general terms, a virtual laboratory is a computer-based activity where students interact with an experimental apparatus or other activity via a computer interface. Typical examples which come to mind include a simulation of an experiment, whereby a student interacts with programmed-in behaviours, and a remote-controlled experiment where a student interacts with real apparatus via a computer link, yet the student is remote from that apparatus. We should distinguish the latter case from a computer-controlled experiment, where a student will directly control an apparatus in his or her vicinity via a computer interface.

Sustainability

A typical laboratory currently uses five times as much energy and water per square foot as a typical office building. Research laboratories are so energy-demanding for a variety of reasons:
• They contain large numbers of containment and exhaust devices.
• They house a great deal of heat-generating equipment.
• Scientists require 24-hour access.
• Irreplaceable experiments require fail-safe redundant backup systems and uninterrupted power supply (UPS) or emergency power.
• Research facilities have intensive ventilation requirements – including "once through" air – and must meet other health and safety codes, which add to energy use.
Key aspects of sustainable design are as follows:
• Increased energy conservation and efficiency.
• Reduction or elimination of harmful substances and waste.
• Improvements to the interior and exterior environments, leading to increased productivity.
• Efficient use of materials and resources.
• Recycling and increased use of products with recycled content.

Cases of the Trend of Laboratory Design

1

2

3

The building is designed with asymmetrical structures and the designers focus on terraces to make many rest spaces

| Korean Institute for Archaeology & Environment

Location:
Chungchongnamdo, Korea
Architect:
Hohyun Park + Hyunjoo Kim
Site Area:
42,980 m²
Completion Date:
2010
Photographer:
Jungmin Seok

- The asymmetrical structures ensure the divisions of different function areas.
- Two staircases were required by local law and U-shaped staircase is planned as a major vertical circulation and a straight run staircase, which connect lobby to research office on the 1st & 2nd floor, is designed for more efficient moving.
- What the designers focused on during design was to make many rest spaces. By placing big and small terraces at many places, the research space becomes more efficient.

4

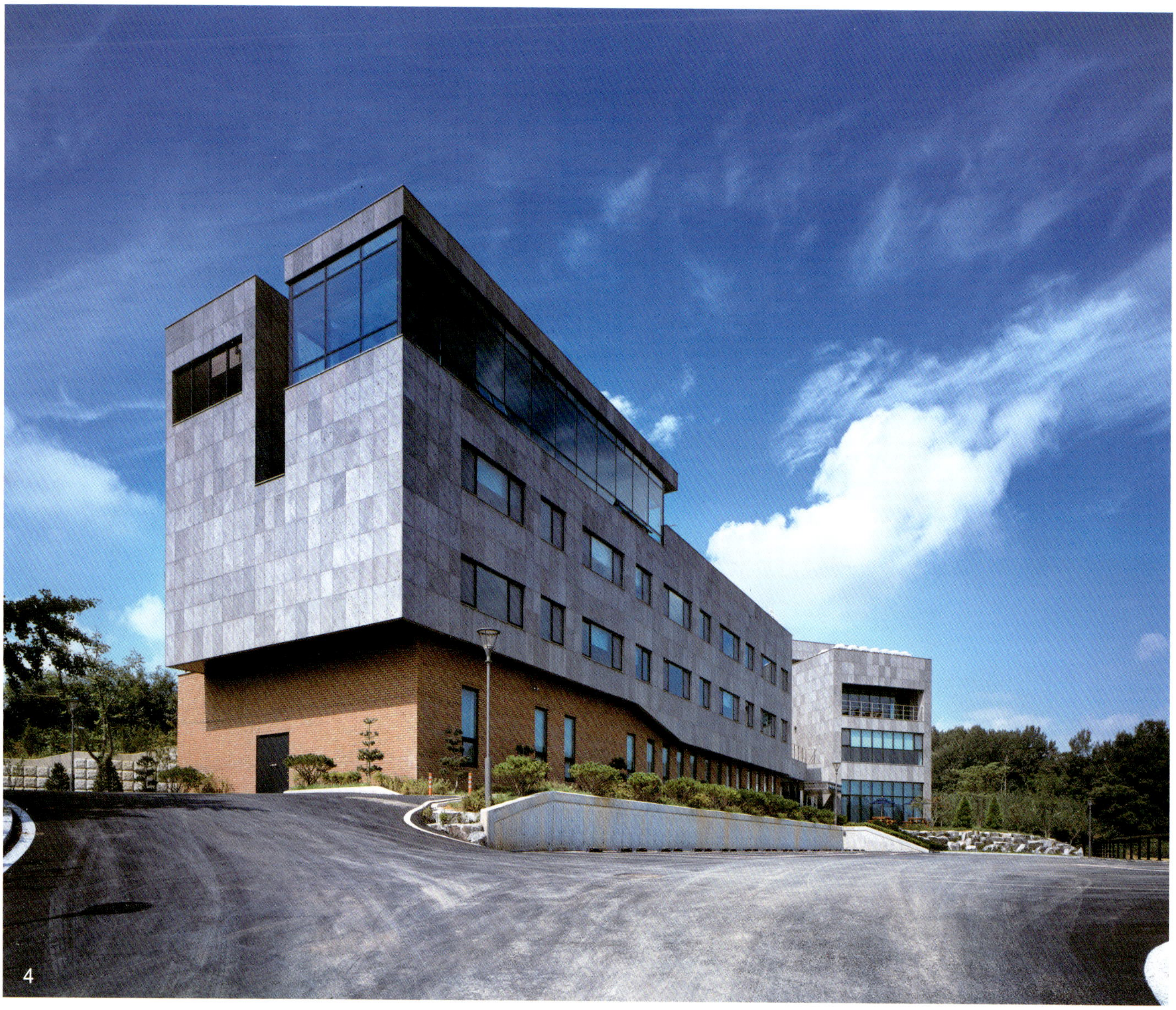

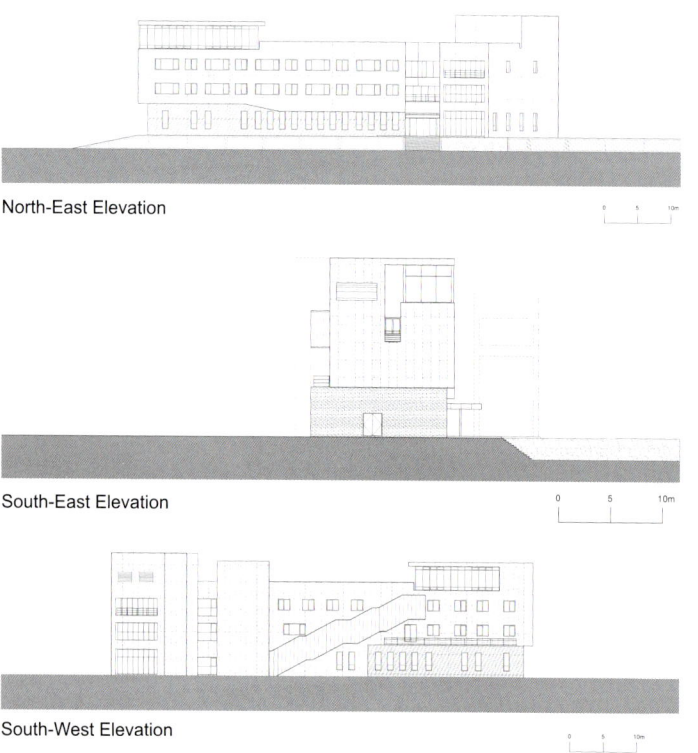

North-East Elevation

South-East Elevation

South-West Elevation

5

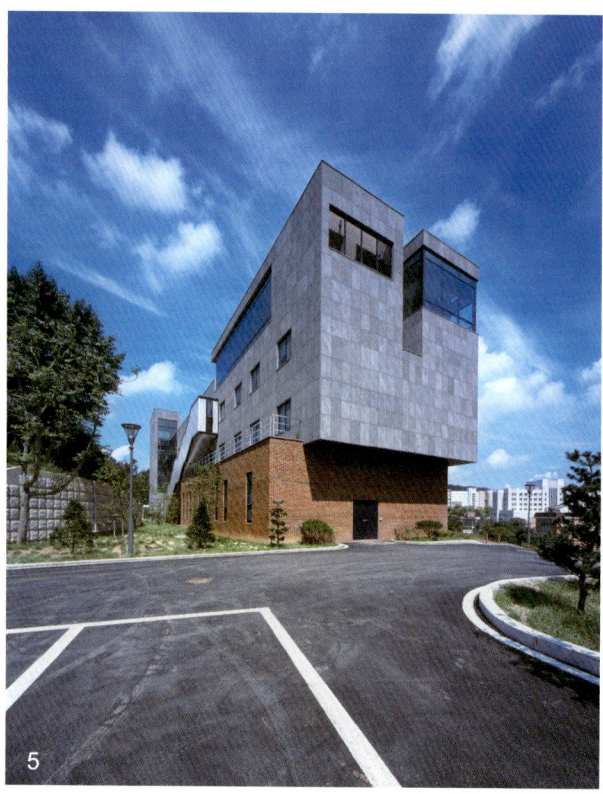

6

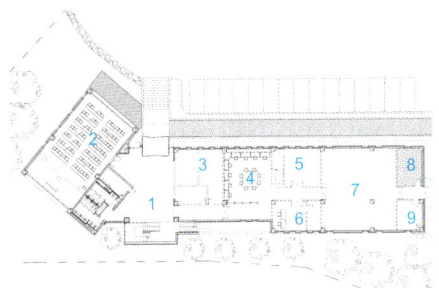

Ground Floor Plan

1. Hall
2. Auditorium
3. Management office
4. Meeting room
5. Office
6. Photographic studio
7. Artifact archive
8. Cleaning room
9. Storage

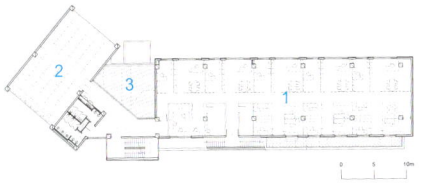

1st Floor Plan

1. Office
2. Library
3. Deck

1. Cafeteria on the 3th floor is accessed through the rooftop garden. Since there are no big buildings around, it has a great view through full glass windows.

2. View of the short wing from southeast

3. The building consists of two asymmetrical wings and a central core. Most of research activities are programmed at the long wing (called research wing), where research offices and storage for artifacts are located. The short wing (called seminar wing) is the space for seminars, library, and meetings.

4. The research wing is divided into a brick finished storage on ground floor and a lava stone finished research offices on the 1st & 2nd floor by shifting of upper mass about 2 metres.

5. An exposed translucent polycarbonate finished staircase and terrace space at the 2nd floor are situated in the gap.

6. The 2nd floor terrace

7. What architects focused on during design was to make many rest spaces. By placing big and small terraces at many places, the research space becomes more efficient.

8. Hall on the 2nd floor

9. Two staircases were required by local law and U-shaped staircase is planned as a major vertical circulation and a straight run staircase, which connect lobby to research office on the 1st & 2nd floor, is designed for more efficient moving.

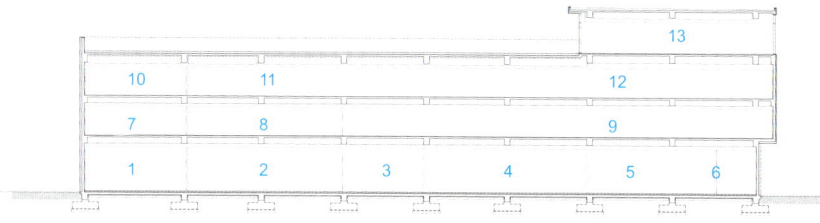

Longitudinal Section

1. Auditorium
2. Lobby
3. Management Office
4. Meeting Room
5. Artifact Archive
6. Cleaning Room
7. Library

8. Hall
9. Research Office
10. Director's Office
11. Hall
12. Research Office
13. Cafeteria

TIPS:

1. Lava stone:

>>> Advantages:

Formed in volcanoes millions of years ago, lava stone tiles and stone mosaics showcase simple beauty. The sophisticated clean textures are the answer to design that requires casualness and elegance.

The beauty of natural stone is a result of inherent and unique variations of colour, shade, texture, and veining. Always blend/mix stone tiles and lay them out prior to installation.

>>> Challenges:

All natural stone must be properly sealed to protect the stone from staining. Sealers may protect the stone with no change in colour or appearance, or you may choose to apply a stone enhancer to deepen the colour and characteristics of the stone.

2. Polycarbonate:

Polycarbonate is a durable material. Although it has high impact-resistance, it has low scratch-resistance. The characteristics of polycarbonate are quite like those of polymethyl methacrylate (PMMA, acrylic), but polycarbonate is stronger, usable in a wider temperature range, yet more expensive. This polymer is highly transparent to visible light, with better light transmission than many kinds of glass.

2

The interior layout encourages cross-disciplinary interaction

| Biological Sciences Complex, University of British Columbia

Location:
Vancouver, Canada
Architect:
Acton Ostry Architects
Construction Area:
15,794 m²
Completion Date:
2011
Photographer:
Acton Ostry Architects

■ A series of seismic buttresses which not only accommodate the latest earthquake structural requirements but are adorned with laminated glass panels upon which are printed abstract forms relating to either the Botany or Zoology Departments.

■ The interior spatial design distinguishes different functions of the labs: Patterned glass, natural wood paneling and additional circulation patterns are used internally to identify the various laboratories and encourage cross-disciplinary interaction.

■ The completed Wings have achieved a LEED Gold standard through a partial building

envelope upgrade, new thermally broken double-glazing, highly efficient lighting, heating and cooling systems, heat recovery, high efficiency pumps and reduced water consumption.

■ Other energy-efficient technologies:

>>Part of a UBC-led pilot project, a prototype technology developed by SunCentral has been introduced which increases the depth that sunshine can reach inside buildings.

>>This groundbreaking solar lighting system reduces the energy demands of the building, therefore lowering greenhouse gas emissions and financial costs to the University.

1. Illuminated at night, the sheer panels are a simple yet highly effective aesthetic and wayfinding device which is used to equal effect within the buildings.

2. One of the most effective introductions is a series of seismic buttresses which not only accommodate the latest earthquake structural requirements but are adorned with laminated glass panels upon which are printed abstract forms relating to either the Botany or Zoology Departments.

3. Patterned glass is used on the exterior façade.

4.5. Patterned glass, natural wood paneling and additional circulation patterns are used internally to identify the various laboratories and encourage cross-disciplinary interaction.

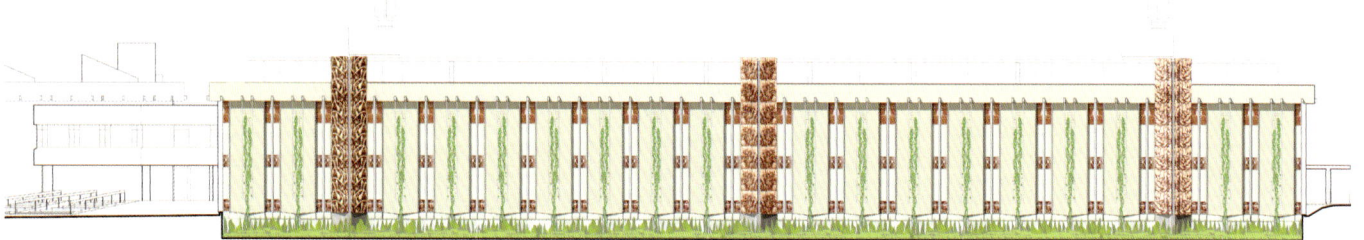

Elevation

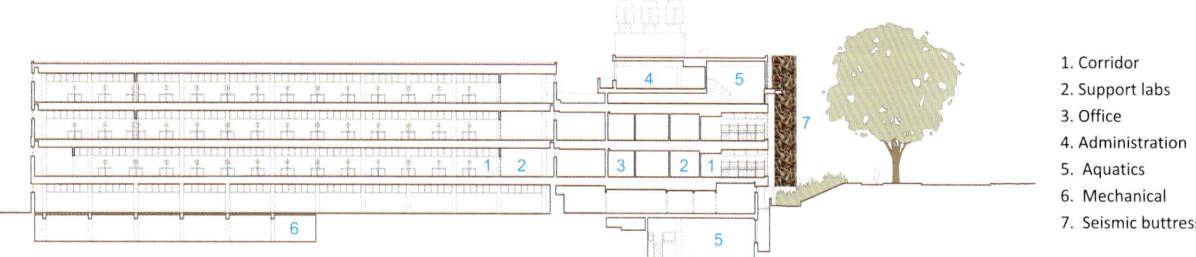

1. Corridor
2. Support labs
3. Office
4. Administration
5. Aquatics
6. Mechanical
7. Seismic buttress

Section

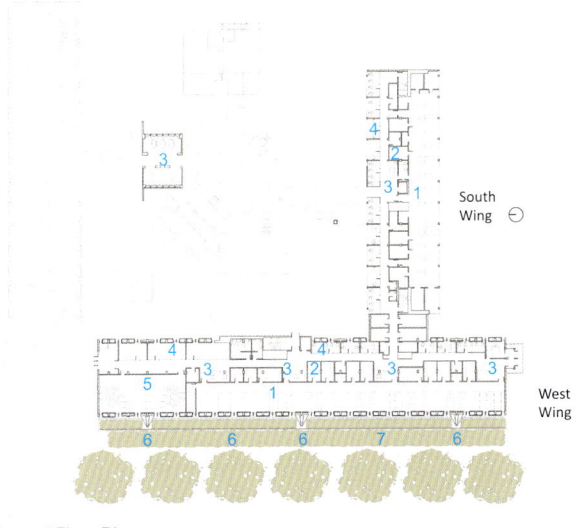

Ground Floor Plan

1. Laboratories
2. Support labs
3. Informal research
4. Office
5. Lecture
6. Seismic buttress
7. Bioswale

South Wing

West Wing

TIPS:

1. Solar lighting system
>>>Benefits
Solar lighting systems save money. Once installed, it costs nothing to power a light with the sun. Adding just one solar-powered light to your home or landscape demonstrates the money savings. Other benefits include reducing your dependence on the planet's limited resources. Solar lights come in both traditional-looking light fixtures and modern-looking fixtures.
>>>LED Lights
Solar-powered lights use LEDs, or light-emitting diode bulbs, which rarely burn out in the lifetime of the light. They emit about 90 percent more light than an incandescent or fluorescent bulb of equal size. They produce almost no heat, making them safer to use in homes and factories.
>>>Drawbacks
Solar lighting has one drawback: it is reliant on the sun's rays. If you use a single solar light as a spotlight, and there is a cloudy period in nature, then you might notice a drop of lighting capability in the spotlight.
2. Laminated glass
>>>Manufacture
There are several laminated glass manufacturing processes:
•Using two or more pieces of glass bonded between one or more pieces of plasticised polyvinyl butyric resin using heat and pressure.
•Using two or more pieces of glass and polycarbonate, bonded together with aliphatic urethane interlayers under heat and pressure.
•Interlaid with a cured resin.
Each manufacturing process may include glass lites of equal or unequal thickness.
>>>Cutting
Plastic interlayers in laminated glass make its cutting difficult. There is an unsafe practice of cutting both sides separately, pouring a flammable liquid such as denatured alcohol into the crack, and igniting it to melt the interlayer to separate the pieces. The following safer methods are recommended:
Special purpose laminated cutting tables
Vertically-inclined saw frames
A blowlamp or hot air blower

6. Zoology laboratory
7. Botany laboratory
8. Classroom
9. Specialty equipment room

1

The use of mimicry creates a kind of sustainable space
| Groningen Centre for Life Sciences

Location:
Groningen, the Netherlands
Architect:
Rudy Uytenhaak Architectenbureau BV
Construction Area:
36,000 m²
Completion Date:
2010
Photographer:
Marcel van der Burg

- Mimicry scores high on sustainability criteria. The building is sustainable in terms of materials used and energy consumption.
- In the building form, the designers put emphasis on visually open space, rather than the form itself.
- The interior organisation features: interaction, dynamism, flexibility

>>Interaction

The upper part of the building is zoned, with laboratories and offices located in two facing areas.

>> Flexibility

The efficiency and flexibility of this model are enhanced by the addition of a third zone incorporating a range of ancillary areas.

>> Vertical circuits (dynamism)

The mix of laboratories, ancillary areas and offices in combination with the spatial and functional qualities of the circuits results in a practical, light and dynamic whole that is highly efficient in both architectural and technical terms.

In this way Mimicry expresses the relationship between form and context, the building and the campus, the landscape and the water, hospitality and collaboration.

- A sustainable building

>> Ample space with relatively small exterior walls

>> The techniques used to create a sustainable technical infrastructure:

Green roof

High-temperature cooling and low-temperature heating using concrete core activation connected to heat and cold storage facilities.

>>Particular attention has been paid to the development of a light, low-maintenance façade (saving construction costs) made up of innovative, façade-length prefabricated polyester wall elements with an especially high insulation value.

The use of these techniques results in an EPC of 0.662: an exceptionally low value for a design dating from 2004.

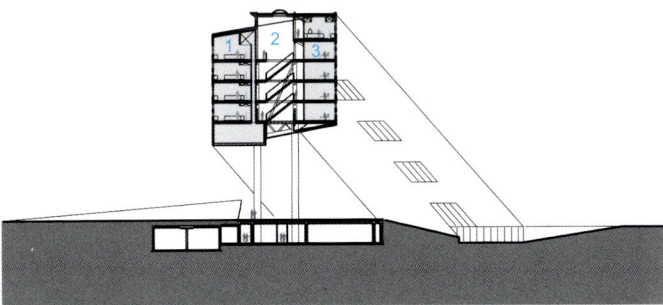

Cross Section

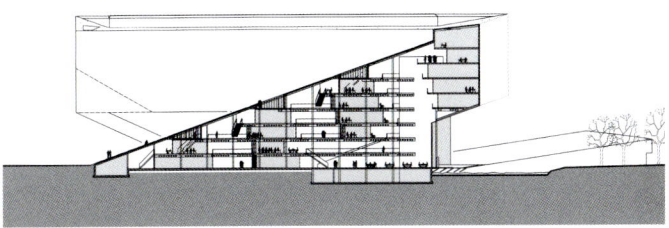

Longitudinal Section

1. Lab zone
2. Intermediate strip (support factions, open areas)
3. Office zone

1. Particular attention has been paid to the development of a light, low-maintenance façade (saving construction costs) made up of innovative, façade-length prefabricated polyester wall elements with an especially high insulation value.
2. The techniques used to create a sustainable technical infrastructure include a green roof, as well as high-temperature cooling and low-temperature heating using concrete core activation connected to heat and cold storage facilities.
3. The volume can be interpreted as a body that rises from the ground and partly vanishes in the perspective and the sloping ground level. The sightlines accentuate the open space rather than the mass of the building.
4. The three research fields are housed in two wings and a bridge that together form the upper part of the building.

5

5. Central hall
6. Because "corridors" are important in places where people work together, it is beneficial to construct the building in such a way that users can traverse it through a variety of circuits.
7. Open area in bridge building
8. Light from above through the open areas and views of the sky, ground or water connect the interiors of the corridors with the outside world.
9. Passage along the offices in bridge building

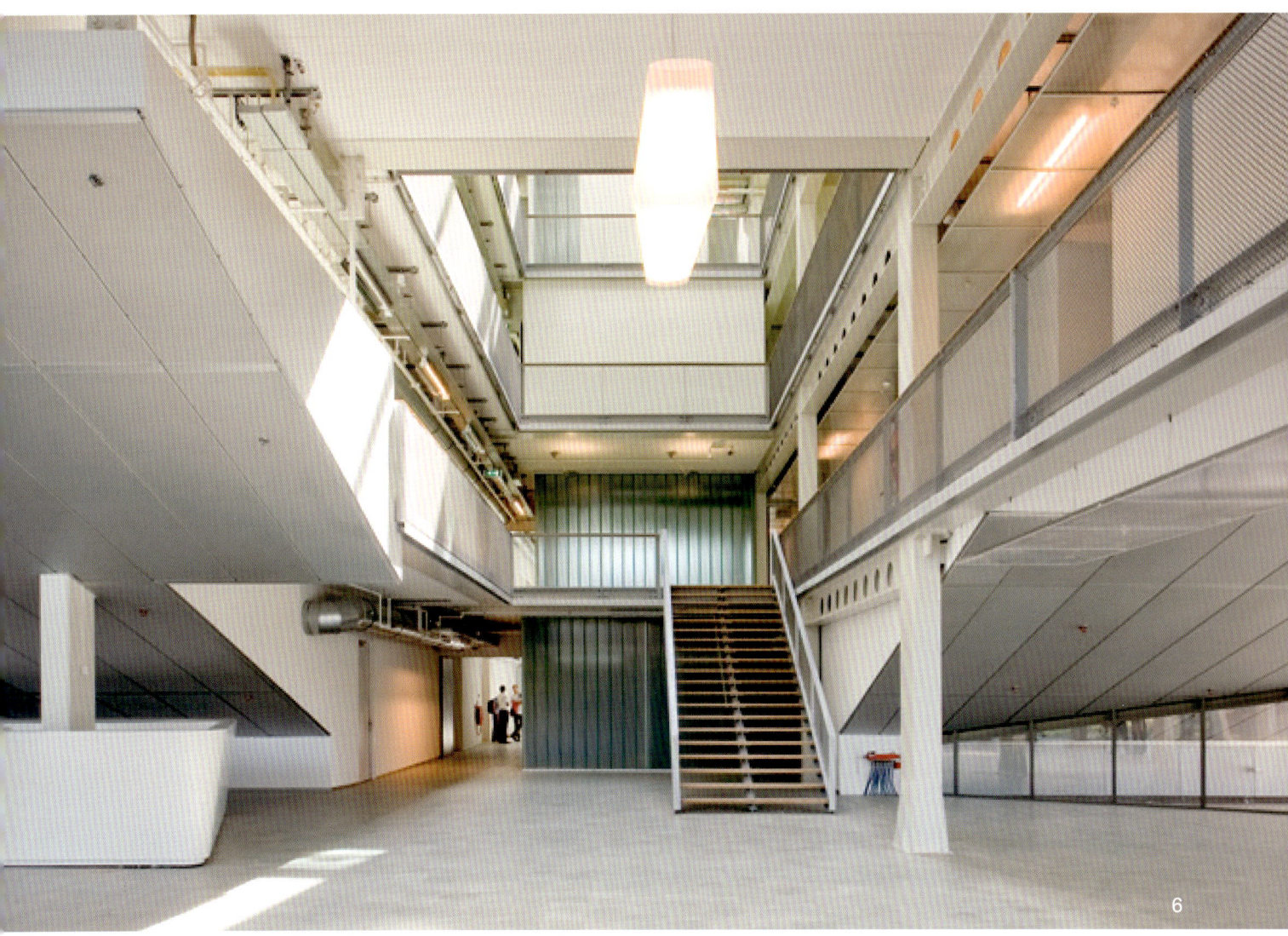

6

TIPS:

1. EPC

EPC is an acronym which stands for Engineering, Procurement and Construction. It is a common form of contracting arrangement within the construction industry.

2. Prefabricated polyester wall

>>>Advantages of prefabrication

Self-supporting ready-made components are used, so the need for formwork, shuttering and scaffolding is greatly reduced.

Construction time is reduced and buildings are completed sooner, allowing an earlier return of the capital invested.

On-site construction and congestion is minimised.

Quality control can be easier in a factory assembly line setting than a construction site setting.

Prefabrication can be located where skilled labour is more readily available and costs of labour, power, materials, space and overheads are lower.

Time spent in bad weather or hazardous

environments at the construction site is minimised.

Less waste may occur.

Advanced materials such as sandwich-structured composite can be easily used, improving thermal and sound insulation and air tightness.

>>> Disadvantages of prefabrication

Careful handling of prefabricated components such as concrete panels or steel and glass panels is required.

Attention has to be paid to the strength and corrosion-resistance of the joining of prefabricated sections to avoid failure of the joint.

Leaks can form at joints in prefabricated components.

Transportation costs may be higher for voluminous prefabricated sections than for the materials of which they are made, which can often be packed more efficiently.

Large prefabricated sections require heavy-duty cranes and precision measurement and handling to place in position.

Larger groups of buildings from the same type of prefabricated elements tend to look drab and monotonous.

Local jobs may be lost, the work done to fabricate the components being located in a place far away from the place of construction. This means that there are less locals working on any construction project at any time, because fabrication is outsourced.

7

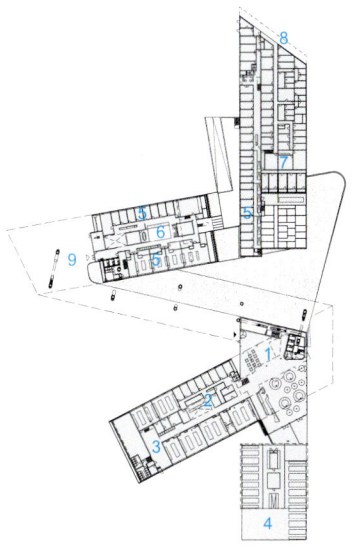

Ground Floor Plan

1. Central hall
2. Teaching and practical halls
3. Technical
4. Greenhouses
5. Office zone
6. Intermediate strip
7. Animal facilities
8. Outdoor animal housing
9. Goods

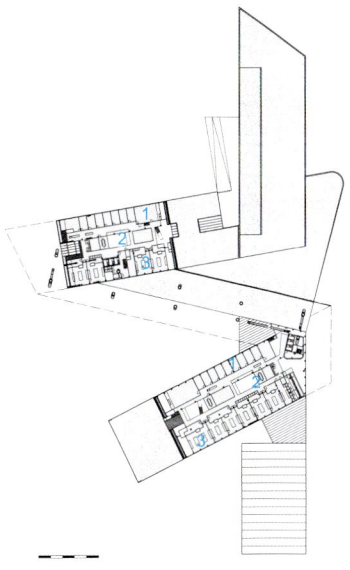

1st Floor Plan

1. Office zone
2. Intermediate strip (support functions, open areas)
3. Lab zone

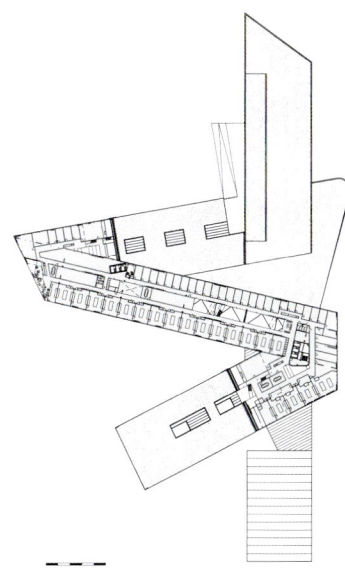

4th Floor Plan

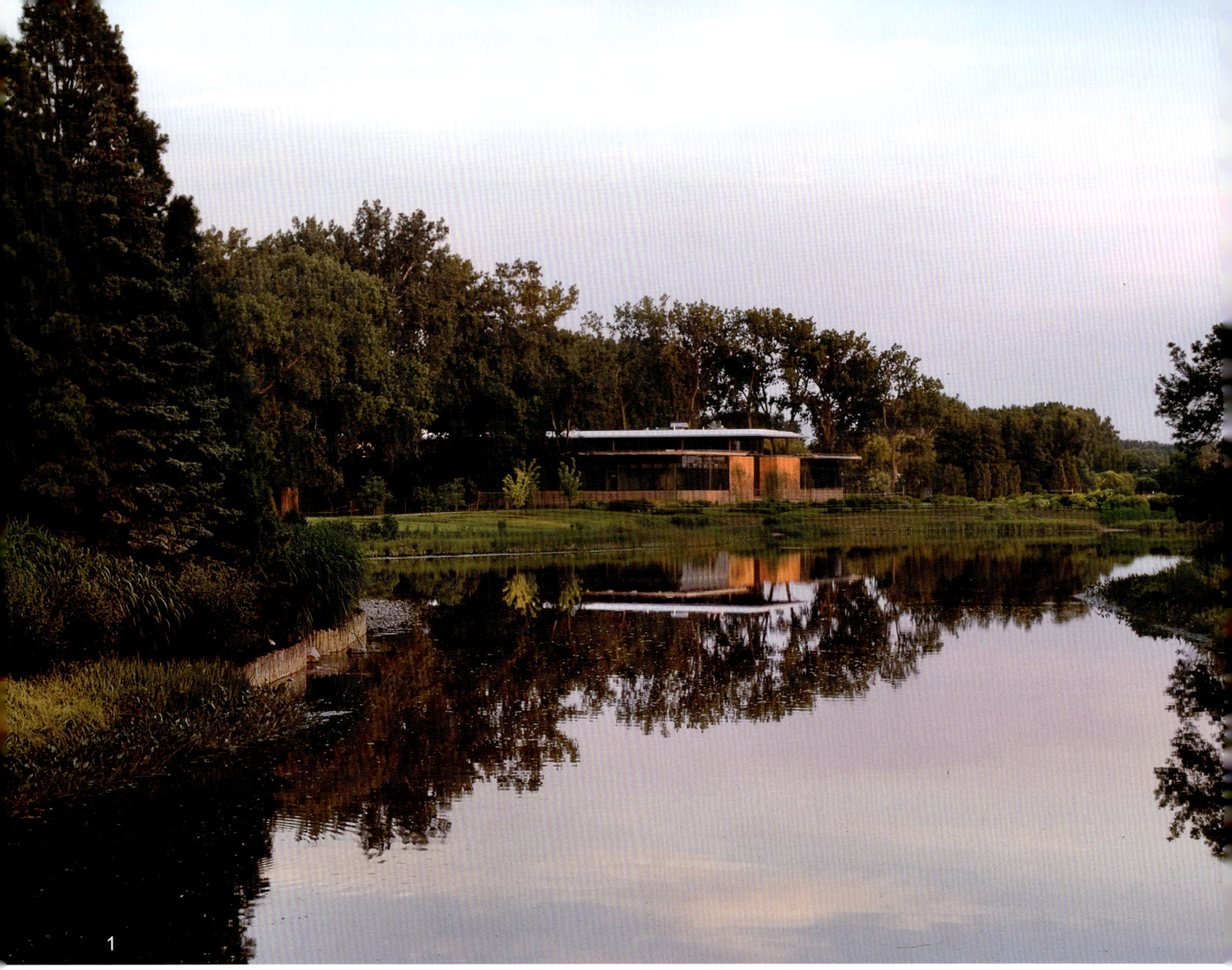
1

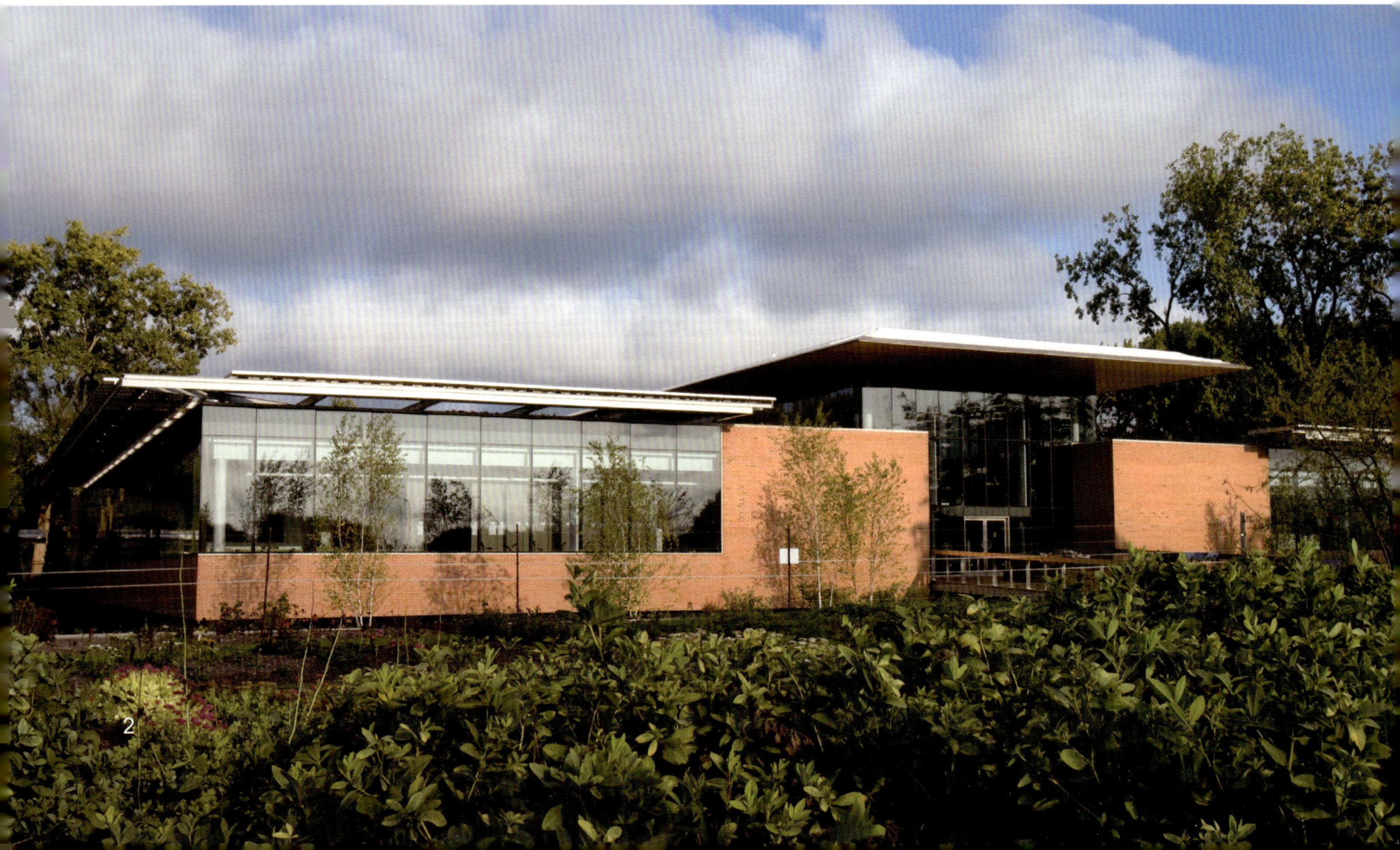
2

The new materials achieve the sustainable design concept

| Chicago Botanic Garden – Plant Conservation Science Centre

Location:
Glencoe, Illinois, USA
Architect:
Booth Hansen
Construction Area:
3,500 m²
Completion Date:
2009
Photographer:
Michelle Litvin

■ Further visual connections are created by the careful use of glazing in the building.
■ The Plant Science Centre is situated on 4.5-foot stilts because it was built on a floodplain. So if the area does flood, there will be no water damage to the building.
■ The botanic garden shows its efforts to sustainable architecture and development in several aspects:
•An energy efficient envelope
•Green Roof Gardens
•The low-e glass for the front façade
•Innovative building systems
•Using new method to make the grey counters to avoid waste of materials

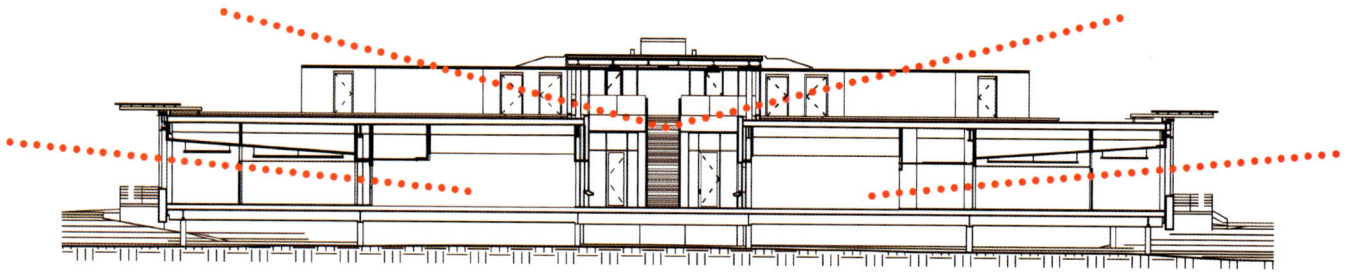

Sun Diagram

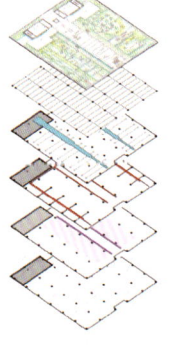

Axonometric Drawing

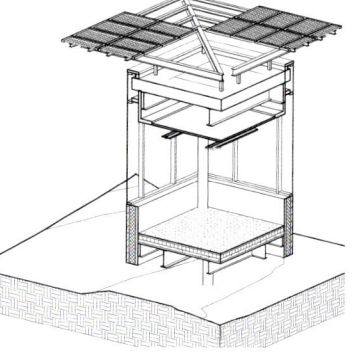

3D Corner

1. Surroundings of the building
2. Front façade, the low-e glass bounces back infrared rays and doesn't heat up. It reflects infrared rays and filters ultraviolet light.
3. The Plant Science Centre is situated on 4.5-foot stilts because it was built on a floodplain. So if the area does flood, there will be no water damage to the building.
4. Upstairs, visitors have access to two Green Roof Gardens where 320 plant varieties grow and solar panels turn sunshine into electricity.
5. The semi-intensive media is a gravel-like soilless mix formulated specifically for roof gardens. It is lighter weight and more porous than soil, allowing water to drain quickly, thus reducing the weight load on the roof.
6. Front entrance
7. Visitors in atrium examining science displays and viewing laboratories. The wood in the ceilings is hemlock from mid-western forests. The ceiling wood needed to flow from the inside of the building to the outside.
8. For the scientists doing research in the building, a sense of community and collaboration is fostered by allowing laboratories to be viewed from each other and the offices.
9. The grey counters are made of a recycled powdered slate cast into a "flowable binder" (like epoxy). Slate was neither milled nor cut to make these counters; instead, the counters contain scraps and ground slate that was poured into a mold until it hardened. This is a responsible way of building counters, creating no waste slate.
10. Plant science library with panoramic views

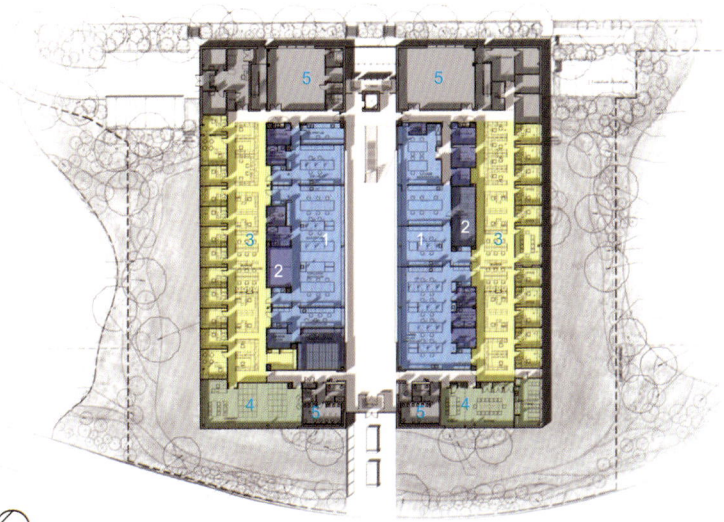

Ground Floor Plan

1. Laboratories
2. Specialty equipment & storage
3. Offices
4. Library & seminar
5. Service & mechanical

TIPS:

1. Green Roof Gardens
>>> Environmental benefits:
 Green roofs are used to:
• Reduce heating (by adding mass and thermal resistance value)
• Reduce cooling (by evaporative cooling) loads on a building by fifty to ninety percent, especially if it is glassed in so as to act as a terrarium and passive solar heat reservoir – a concentration of green roofs in an urban area can even reduce the city's average temperatures during the summer
• Reduce stormwater run off
• Natural Habitat Creation
• Filter pollutants and carbon dioxide out of the air which helps lower disease rates such as asthma
• Filter pollutants and heavy metals out of rainwater
• Help to insulate a building for sound; the soil helps to block lower frequencies and the plants block higher frequencies
• If installed correctly, many living roofs can contribute to LEED points
• Increase agricultural space
• With green roofs, water is stored by the substrate and then taken up by the plants from where it is returned to the atmosphere through transpiration and evaporation
• Green roofs not only retain rainwater, but also moderate the temperature of the water and act as natural filters for any of the water that happens to run off
>>>Financial benefits
• Increase roof life span dramatically
• Increase real estate value
• Reduction in energy usage
The low-e glass:
>>> Definition:
Low-e glass is a type of treated glass that conducts visible light while controlling the passage of heat.
>>> Theory:
This glass works by reflecting heat back to its source. All objects and people give off varying forms of energy, affecting the temperature of a space. Long wave radiation energy is heat, and short wave radiation energy is visible light from the sun. The coating used to make low-e glass works to transmit short wave energy, allowing light in, while reflecting long wave energy to keep heat in the desired location.
>>> Application:
Low-e glass comes in high, moderate and low gain panels. In especially cold climates, heat is preserved and reflected back into a house to keep it warm. This is accomplished with high solar gain panels. In especially hot climates, low solar gain panels work to reject excess heat by reflecting it back outside the space. Moderate solar gain panels are also available for areas with temperature fluctuations.
>>> Challenge:
Visibility was a problem with some of the first low-e glasses, as original panes were said to have a brownish tint. Technology and manufacturing has continued to improve its quality resulting in a spectrally selective glass that allows the best possible visibility while still filtering heat.

10

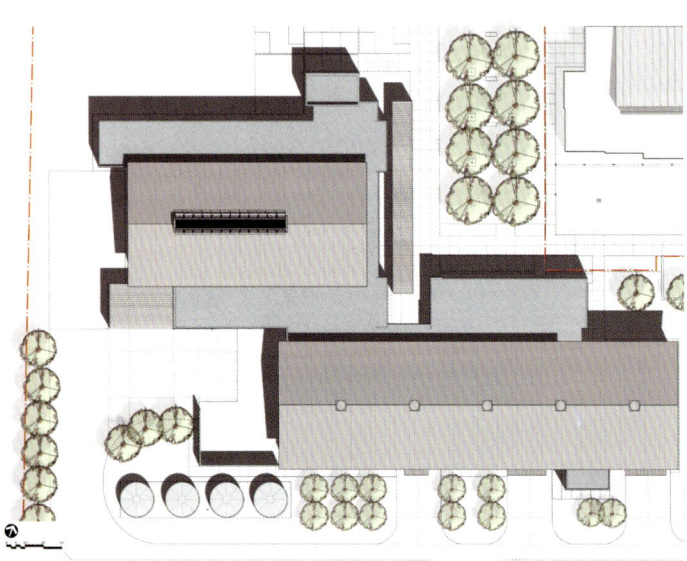

Site Plan

Ground Floor Plan

Demonstration on energy efficiency and strong sustainability

| Teaching and Research Winery and the August A. Busch III Brewing and Food Science Laboratory

Location:
Davis, California, USA

Architect:
Flad Architects/Andrew Cunningham/
Stevens Williams

Construction Area:
10,104 m²

Completion Date:
2010

Photographer:
Robert Canfield

▪ Overall design concept: It is a living model, where the effectiveness of energy efficient technologies is directly monitored and demonstrated in a highly sustainable setting.

▪ Achievements in green and sustainable design:

This is the first LEED Platinum building on this campus and the third in the UC system.

It houses the world's first LEED Platinum winery.

First LEED Platinum brewery.

First LEED Platinum food-processing pilot plant.

It is one of about a half dozen laboratories and the first process science building to attain this level of performance.

▪ This project is entirely donor-funded. And platinum performance required specific innovations that engaged the vision of industry donors and addressed the demands of an arid climate. These features include:

▪ Capture and storage of 176,000 gallons of rainwater satisfies annual irrigation requirements and all non-potable demands. This is the first large-scale rainwater harvesting system in the region to accommodate both uses.

▪ A Clean-in-Place (CIP) system similar to that used in pharmaceutical manufacture collects, treats, and re-uses all cleaning water, reducing demand by about 80 percent.

▪ Research fermentors are piped for CO_2 capture, allowing later conversion to solid state.

▪ The building is a total learning environment. Real time data on metering of building systems is displayed.

1. Construction of the WBF Lab utilises a variety of sustainable materials, including recycled glass, repurposed wood for interior paneling, and FSC-certified lumber.

2. The complex includes facilities dealing with the production of wine, beer, and traditional food processing in addition to necessary ancillary space and learning laboratories characteristic of a traditional academic structure.

3. Rooftop photovoltaic cells provide all of the facility's power at peak load.

4. Main entrance at night

5. Courtyard

6. One environmentally responsible features is maximum use of natural light.

7. The new sustainable winery building will make it possible to sequester the captured carbon dioxide so that it will not contribute to global warming. The new food-processing equipment can minimises energy and water requirements, use of recycled glass in the flooring.

8. Interior paneling recycled from a 1928 wooden aqueduct, and use of lumber harvested from sustainably certified forest operations.

9. Corridor

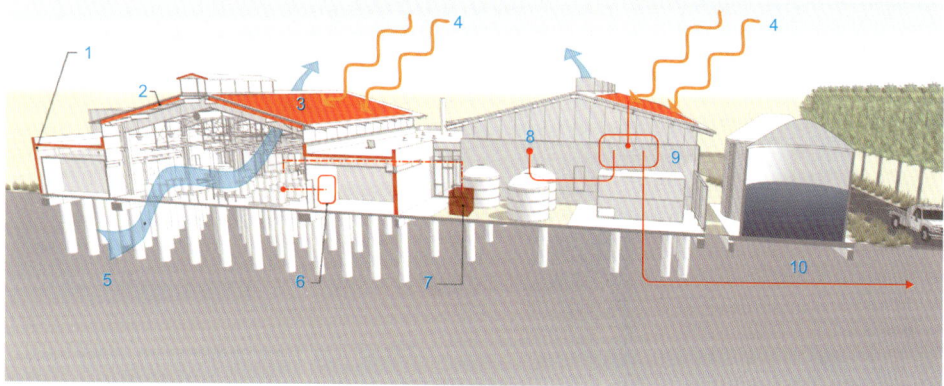

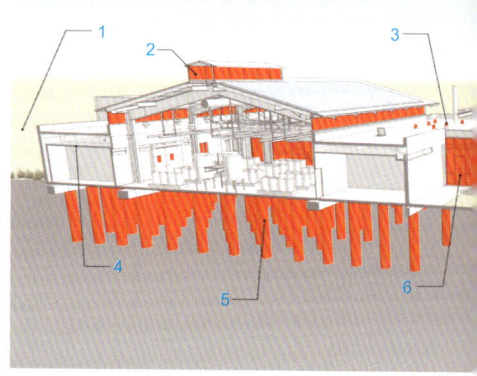

Energy

1. High performance wall insulation
2. High performance roof insulation
3. Photovoltaic
4. Sun's rays
5. Night purge for CO_2 removal
6. Energy metering
7. CO_2 sequestering
8. To building
9. Energy convertor
10. To grid

Environment

1. Percentage of site used
2. Clearstories
3. Solar tubes
4. Recycled glass/wood
5. Geo-piers/Ground stabilisation
6. Views from 92% of spaces
7. Natural site remediation
8. Shading of hard scape
9. Native landscape

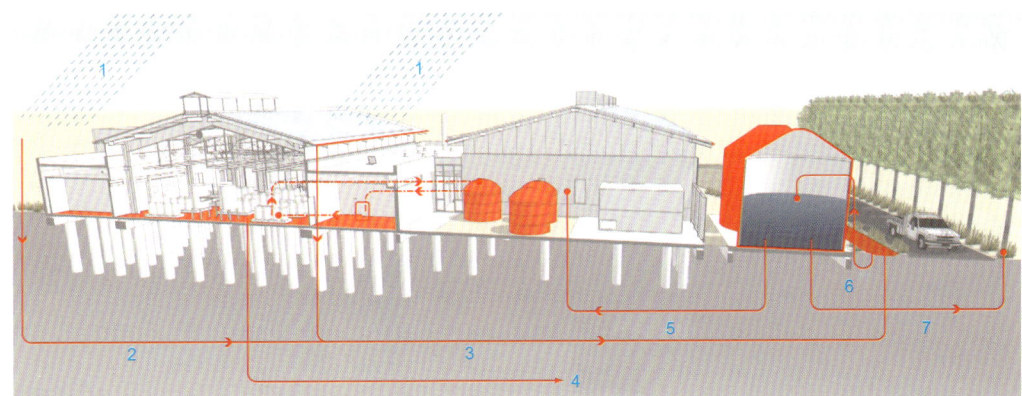

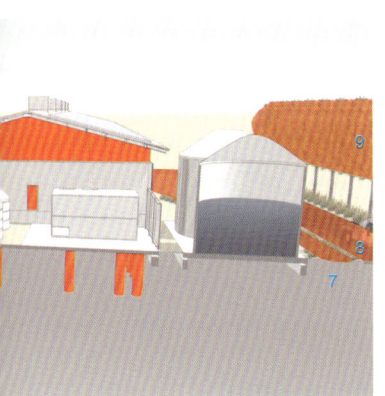

Water

1. Rain
2. Water from site to storage tanks
3. Water from gutters to storage tanks
4. Potential recycling of wash down waste
5. Gray water for flushing toilets
6. Water from bios wale to storage tank
7. Irrigation

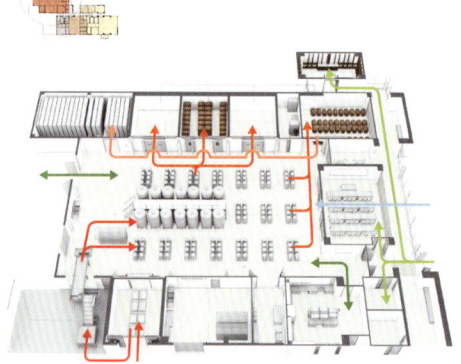

 Process flow
 Product flow
Research flow
Visitor flow
View corridor

Process flow
Product flow
Research flow
Visitor flow
View corridor
CIP

Process flow
Product flow
Research flow
Visitor flow
Equipment

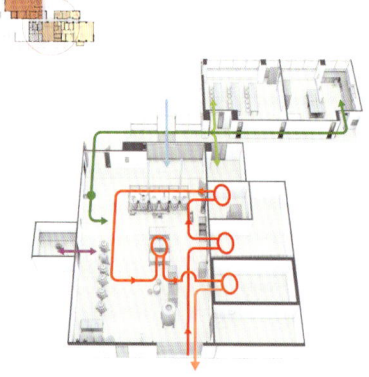

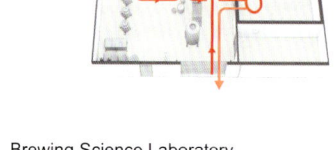

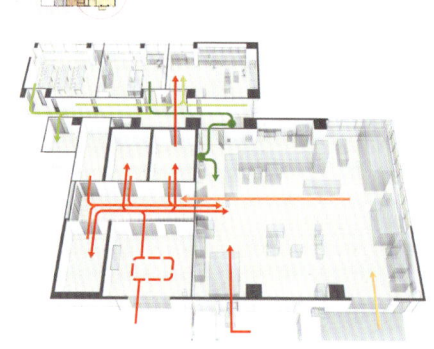

Teaching and Research Winery

Brewing Science Laboratory

Food Science Laboratory

TIPS:

1. Clean-in-Place (CIP) system:

Clean-in-Place (CIP) is a method of cleaning the interior surfaces of pipes, vessels, process equipment, filters and associated fittings, without disassembly.

Depending on soil load and process geometry, the CIP design principle is one of the following:

• Deliver highly turbulent, high flow-rate solution to effect good cleaning (applies to pipe circuits and some filled equipment).

• Deliver solution as a low-energy spray to fully wet the surface (applies to lightly soiled vessels where a static sprayball may be used).

• Deliver a high energy impinging spray (applies to highly soiled or large diameter vessels where a dynamic spray device may be used).

Elevated temperature and chemical detergents are often employed to enhance cleaning effectiveness.

2. Awards Name:

LEED Platinum

R&D Magazine Lab of the Year High Honors

2010 "Best of California" Construction Award

AIA San Francisco Merit Award for Energy & Sustainability

ASLA San Francisco Honour Award – Best in Category

ASLA San Francisco President's Award

2011 California Higher Education, Energy & Sustainability Best Practice Award for Water Efficiency

Western Pacific Region DBIA Best Project – Educational

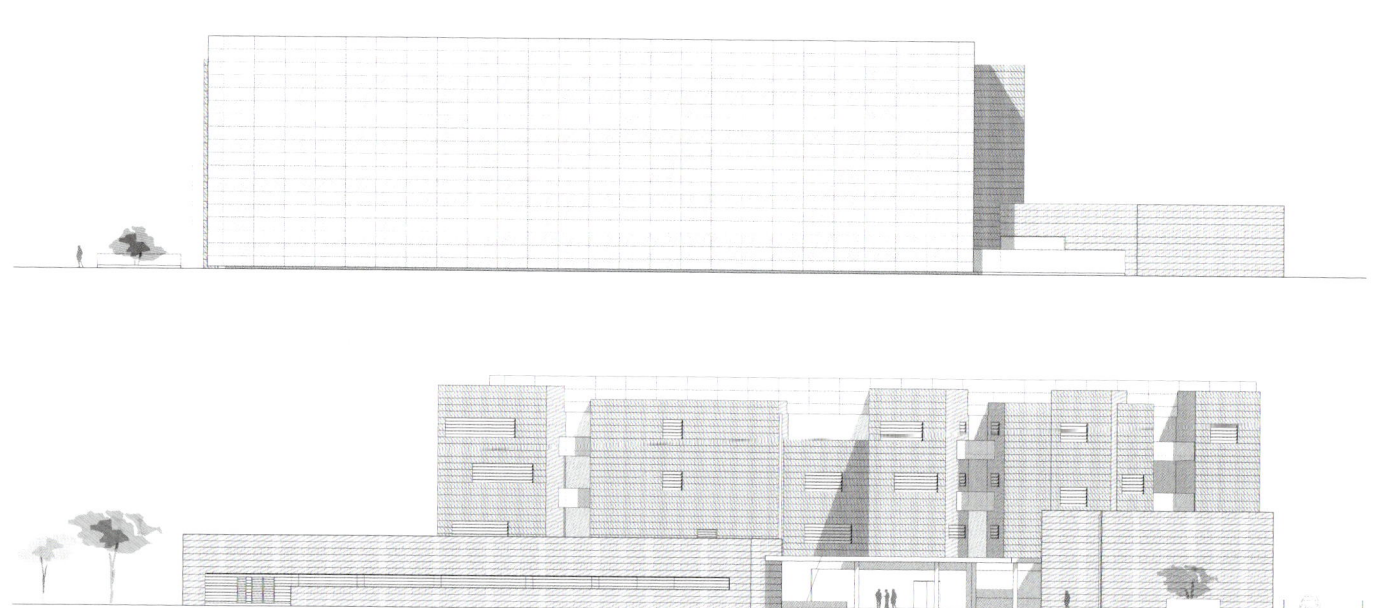

Elevations

The façades are designed with insulating materials and the photovoltaic panels provide clean energy advantage

| GENyO Centre for Genomics and Oncological Research

Location:
Málage, Spain
Architect:
Planho Consultores
Construction Area:
6,500 m²
Completion Date:
2010
Photographer:
Alejandro González

▪ The design of façades that simultaneously isolates and clean energy advantage by integrating photovoltaic panels.

▪ The available land and the functional use of the building led the designers to think about a high-rise building. The position, aspect, orientation, direction and shape of the plot of land were determining factors in its implementation and planning.

▪ Walking through the inner corridor with every single wall being planned and executed to reflect the vision of conceptual duality, one will participate in what is consolidated and what is in the future, in knowledge and what we have yet to learn.

1. The north facing façade, designed with an irregular structure, consistent with its views of the consolidated city, will house all departments to support research units.

2. The coating of the building reflects the proposed functional concept.

3. Main entrance

4. The façade of the southeast-oriented linear block, designed in a very rational form, can enjoy the views of the Sierra Nevada and the plain of Granada.

5. The design of façades that simultaneously isolates and clean energy advantage by integrating photovoltaic panels

6. Courtyard

7. Entrance hall

8. Walking through the inner corridor with every single wall being planned and executed to reflect the vision of conceptual duality, one will participate in what is consolidated and what is in the future, in knowledge and what we have yet to learn.

9. Emergency shower

10. Laboratory

11. Auditorium

Laboratory & Test Cabinet
Pre-test Room
Ancillary Facility
Administration Room
Lobby

Rest Room
Training Room
Electrical Distribution Facility
Foyer

1st Floor/2nd Floor/3rd Floor Plan

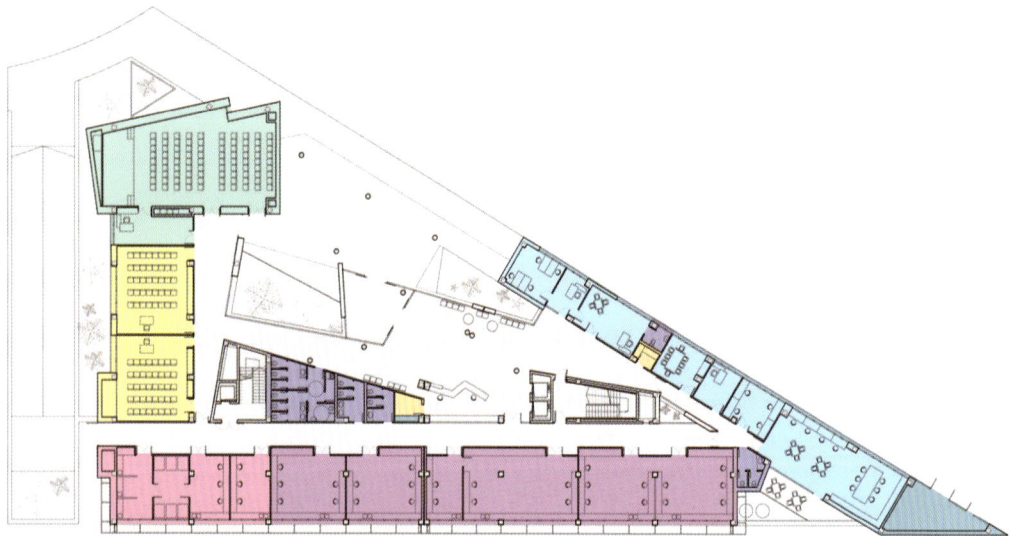

Ground Floor Plan

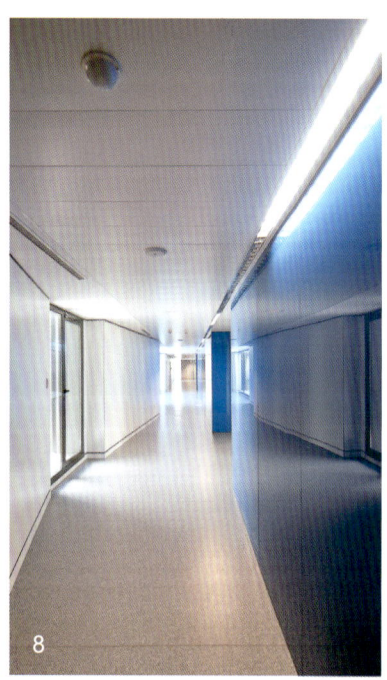

10

11

TIPS:

1. Photovoltaic glass:

>>>Definition

PV glass is a special glass that is used to generate PV power. The solar cells are embedded between two glass panes and, when used in PV glass application, they are either crystalline silicon or thin film.

>>>Application

Photovoltaic glass is a special glass with integrated solar cells, to convert solar energy into electricity. This means that the power for an entire building can be produced within the roof and façade areas. The solar cells are embedded between two glass panes and a special resin is filled between the panes, securely wrapping the solar cells on all sides. Each individual cell has two electrical connections, which are linked to other cells in the module, to form a system which generates a direct electrical current.

>>>Challenges

The current application of PV glass technology is mainly to power the building. Due to limited availability of sunlight, PV glass is not in any position to be able to provide sufficient energy for the whole building. Yet, it is believed that there is still room for growth in this particular industry as long as the power generated is enough to power basic home appliances like cell phone recharge and computers.

2. Corrugated aluminium sheets:

Corrugated galvanised iron (colloquially corrugated iron, commonly abbreviated CGI) is a building material composed of sheets of hot-dip galvanised mild steel, cold-rolled to produce a linear corrugated pattern in them. The corrugations increase the bending strength of the sheet in the direction perpendicular to the corrugations, but not parallel to them. Normally each sheet is manufactured longer in its strong direction.

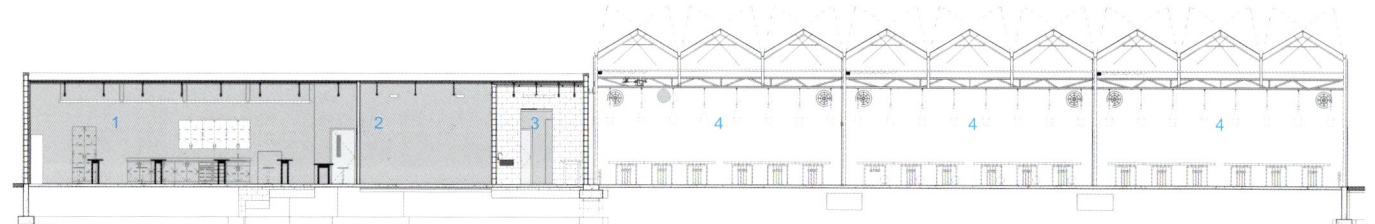

1. Multipurpose
2. Mechanical
3. Head house
4. Greenhouse

Section

Utilisation of efficient water-conserving facilities and green house technology
| Joliet Junior College Greenhouse Facility

Location:
Joliet, Illinois, USA
Architect:
Legat Architects, Inc.
Construction Area:
1,107 m²
Completion Date:
2010
Photographer:
Jim Steinkamp/Steinkamp Photography

• Rainwater harvesting system:
The facility reduces water consumption by 40% over the design baseline. A three-tiered rainwater harvesting system draws first from rainwater stored in underground tanks, then from non-potable well water, then (if needed) from city water. The tanks retain 13,600 gallons of captured or non-potable water for greenhouse irrigation. An on-site sanitary treatment system will support outdoor native vegetation.
Other water-conserving features include:
>> Drip irrigation
>> "Ebb and flow" benches, which continually recycle water used on plants
>> Hydroponics (growing without soil)
>> "Pulse" watering for hanging plants

Design Points:
• A Priva system and a building automation system moderate the environment and water resources in each house. Every six seconds, monitors check internal conditions based on parameters set by the college.
• The facilities and technologies used in the green house:
A black shade cloth system extends night in House 1, while high-intensity discharge lights extend day in House 3. Thermal curtains in each house help retain heat.
• The mechanically-ventilated roof opens and closes to moderate house temperatures, while horizontal air fans at 11 feet help circulate air in each house.

1. Beneath the east façade of the greenhouse facility is the formal garden designed by faculty and students. A path paved with bricks made of ground-up rubber and milk cartons passes through nine gardens with themes ranging from Alpine to Japanese.

2. The facility's north-south orientation enables different levels of daylight and solar heating in each house.

3. Energy reflective roof surface to reflect solar energy and reduce cooling loads

4. An educational building connects to the greenhouses. It includes a multipurpose classroom, a head house, coolers, pesticide storage, and shared office space.

5. The mechanically-ventilated roof opens and closes to moderate house temperatures, while horizontal air fans help circulate air in each house.

6. Classroom

7. Three passively and mechanically ventilated greenhouses provide flexible growing environments.

TIPS:

1. Rainwater harvesting system:
>>> Advantages in urban areas
Rainwater harvesting can ensure an independent water supply during main water restrictions and are effective in "green droughts", capturing low rainfall events where runoff coefficients are not sufficient to cause flow into dam storages. Though somewhat dependent on end-use and maintenance, yields are usually of acceptable quality for most household needs and renewable at acceptable volumes, despite climate change forecasts. It produces beneficial effects by reducing peak storm water runoff and processing costs. Rainwater harvesting systems are simple to install and operate. Running costs are negligible, and they provide water at the point of consumption.

2. Drip irrigation
>>> The advantages of drip irrigation are:
• Fertiliser and nutrient loss is minimised due to localised application and reduced leaching.
• Water application efficiency is high.
• Field levelling is not necessary.
• Fields with irregular shapes are easily accommodated.
• Recycled non-potable water can be safely used.
• Moisture within the root zone can be maintained at field capacity.
• Soil type plays a less important role in frequency of irrigation.

6

- Soil erosion is minimised.
- Weed growth is minimised.
- Water distribution is highly uniform, controlled by output of each nozzle.
- Labour cost is less than other irrigation methods.
- Variation in supply can be regulated by regulating the valves and drippers.
- Fertigation can easily be included with minimal waste of fertilisers.
- Foliage remains dry, reducing the risk of disease.
- Usually operated at lower pressure than other types of pressurised irrigation, reducing energy costs.

>>>The disadvantages of drip irrigation are:
- Expense: initial cost can be more than overhead systems.
- Waste: the sun can affect the tubes used for drip irrigation, shortening their usable life.
- Clogging: if the water is not properly filtered and the equipment not properly maintained, it can result in clogging.
- Drip irrigation might be unsatisfactory if herbicides or top dressed fertilisers need sprinkler irrigation for activation.
- Drip tape causes extra cleanup costs after harvest. Users need to plan for drip tape winding, disposal, recycling or reuse.
- Waste of water, time and harvest, if not installed properly. These systems require careful study of all the relevant factors like land topography, soil, water, crop and agro-climatic conditions, and suitability of drip irrigation system and its components.
- Germination problems: in lighter soils subsurface drip may be unable to wet the soil surface for germination. Requires careful consideration of the installation depth.
- Salinity: most drip systems are designed for high efficiency, meaning little or no leaching fraction. Without sufficient leaching, salts applied with the irrigation water may build up in the root zone, usually at the edge of the wetting pattern. On the other hand, drip irrigation avoids the high capillary potential of traditional surface-

applied irrigation, which can draw salt deposits up from deposits below.

3. Hydroponics (growing without soil)

>>>Advantages

Some of the reasons why hydroponics is being adapted around the world for food production are the following:

• No soil is needed for hydroponics.

• The water stays in the system and can be reused – thus, lower water costs.

• It is possible to control the nutrition levels in their entirety – thus, lower nutrition costs.

• No nutrition pollution is released into the environment because of the controlled system.

• Stable and high yields.

• Pests and diseases are easier to get rid of than in soil because of the container's mobility.

• It is easier to harvest.

• No pesticide damage.

>>>Disadvantages

Without soil as a buffer, any failure to the hydroponic system leads to rapid plant death. Other disadvantages include pathogen attacks such as damp-off due to verticillium wilt caused by the high moisture levels associated with hydroponics and over watering of soil based plants. Also, many hydroponic plants require different fertilisers and containment systems. To produce the mineral wool and the fertilisers that are needed to use this method, a large amount of energy is required.

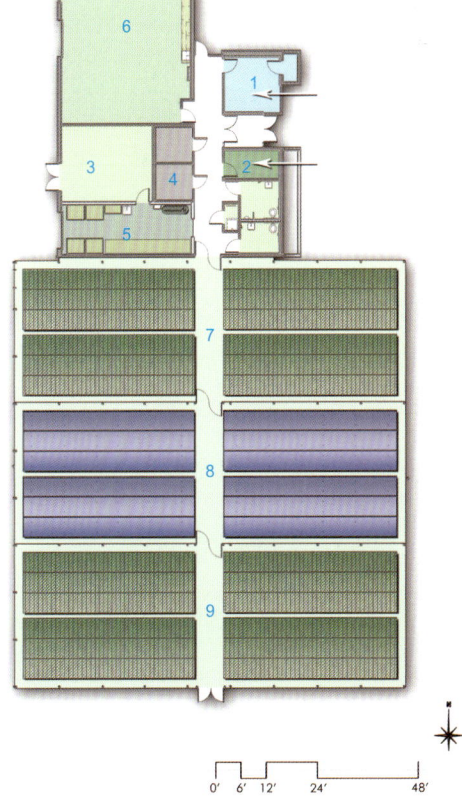

1. Faculty office
2. Pesticide storage
3. Mech.
4. Coolers
5. Head house
6. Multipurpose classroom
7. Greenhouse 1
8. Greenhouse 2
9. Greenhouse 3

Ground Floor Plan

1

The building form interacts closely with the green facility

| Norman Hackerman Building at the University of Texas at Austin

Location:
Austin, USA
Architect:
CO Architects, Taniguchi Architects
Construction Area:
27,871 m²
Completion Date:
2010
Photographer:
Tom Bonner

■ With the development of science and technology, more and more interdisciplinary sciences emerge. With its high requirements in lab environment, this new type of interdisciplinary lab building is gradually playing a unique role in research buildings.

Because of Austin's particular hot and humid climate, the design level dew point for the HVAC systems is set at the demanding levels of 98°F dry bulb and 80 percent relative humidity.

In short, the building had to keep occupants comfortable even while drawing in and cooling immense amounts of hot and humid air. This required outside air to be cooled down to 55°F in order to ensure that enough moisture was drawn out of the air so as not to create humid conditions in the laboratories. The result was a significant amount of chilled condensate water that was later re-used throughout the building.

■ In order to combine the functions of the lab and the local climate, the designers designed a long and narrow building form to support the large cantilever of solar water heater.

■ The façade materials and its features:

The two-storey base façade is made from light buff limestone, similar in hue to the granite on the signature University of Texas Tower and the middle layer combines red-brown brick matching adjacent buildings with glass.

■ In order to fit the strong light, the designers utilise embedded double layer porch to provide two secondary entrances.

■ Green Initiatives

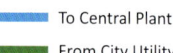

1. High efficiency AHU's
2. Solar hot water
   Heating array
3. Light coloured
   Work surfaces
4. High performance
   Building envelope
5. Recovered water system
   Water system
6. Condensation water
   Re-use
7. Local cooling units
8. Solar hot water supply
9. Reclaimed irrigation

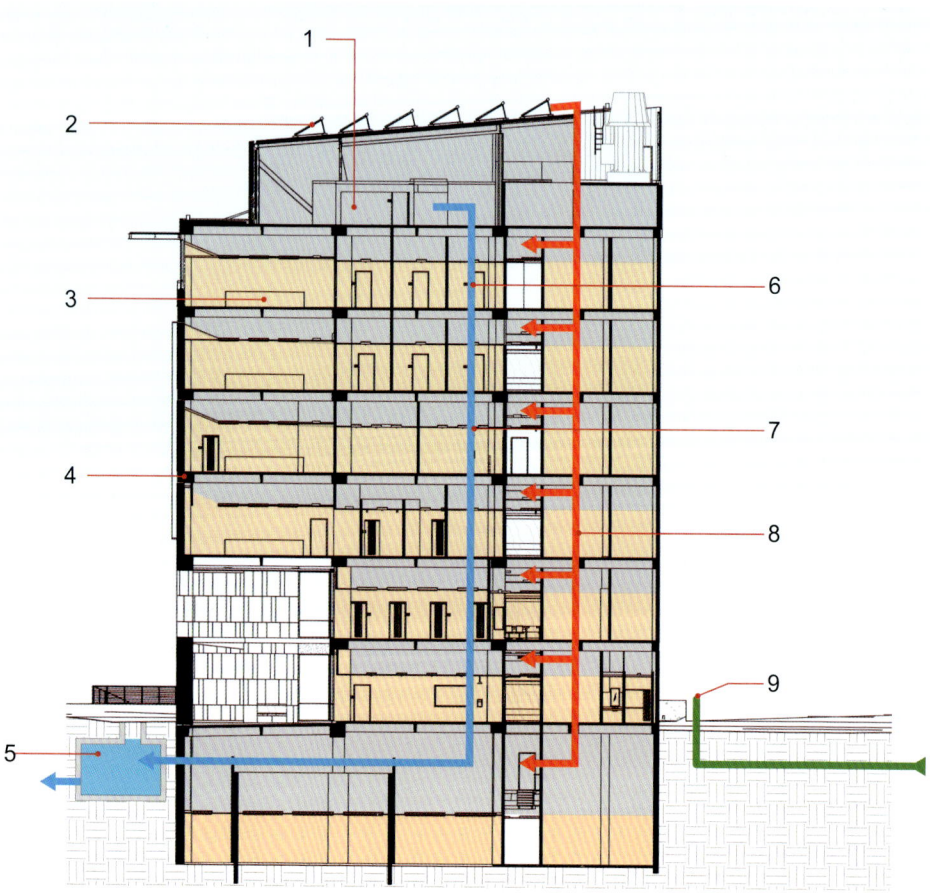

5

>> The building is a case study that would implement the "Labs21" approach.

>> Total energy use is anticipated to be 34-percent below ASHRAE 90.1 requirements.

>> Many other proven strategies were adopted to obtain energy and water savings, and attain a LEED Gold rating –

Variable air-volume HVAC systems

Local re-circulating cooling units for equipment rooms and high heat-load spaces

Cascading air from surrounding spaces into high fume-hood density labs

Variable frequency drives for all air- and water-moving equipment

Photocell sensors to control lights when rooms are unoccupied

Automatic daylight controls

Low-VOC, recycled-content, and local materials

City of Austin "purple pipe" reclaimed-water irrigation system

Storm-water retention system

Daylight harvesting system

Diverting 82% of construction waste from landfill

Although there were challenges in creating ideal lab conditions in a building requiring many air changes in a harsh climate, there were some site-specific advantages, including a central-campus chilled-water distribution system, and a central power and steam-heat system.

2nd Floor Plan

1. Centre for learning and memory research labs
2. Break room
3. School of biological sciences administration
4. Concerence room
5. Centre for learning and memory administration

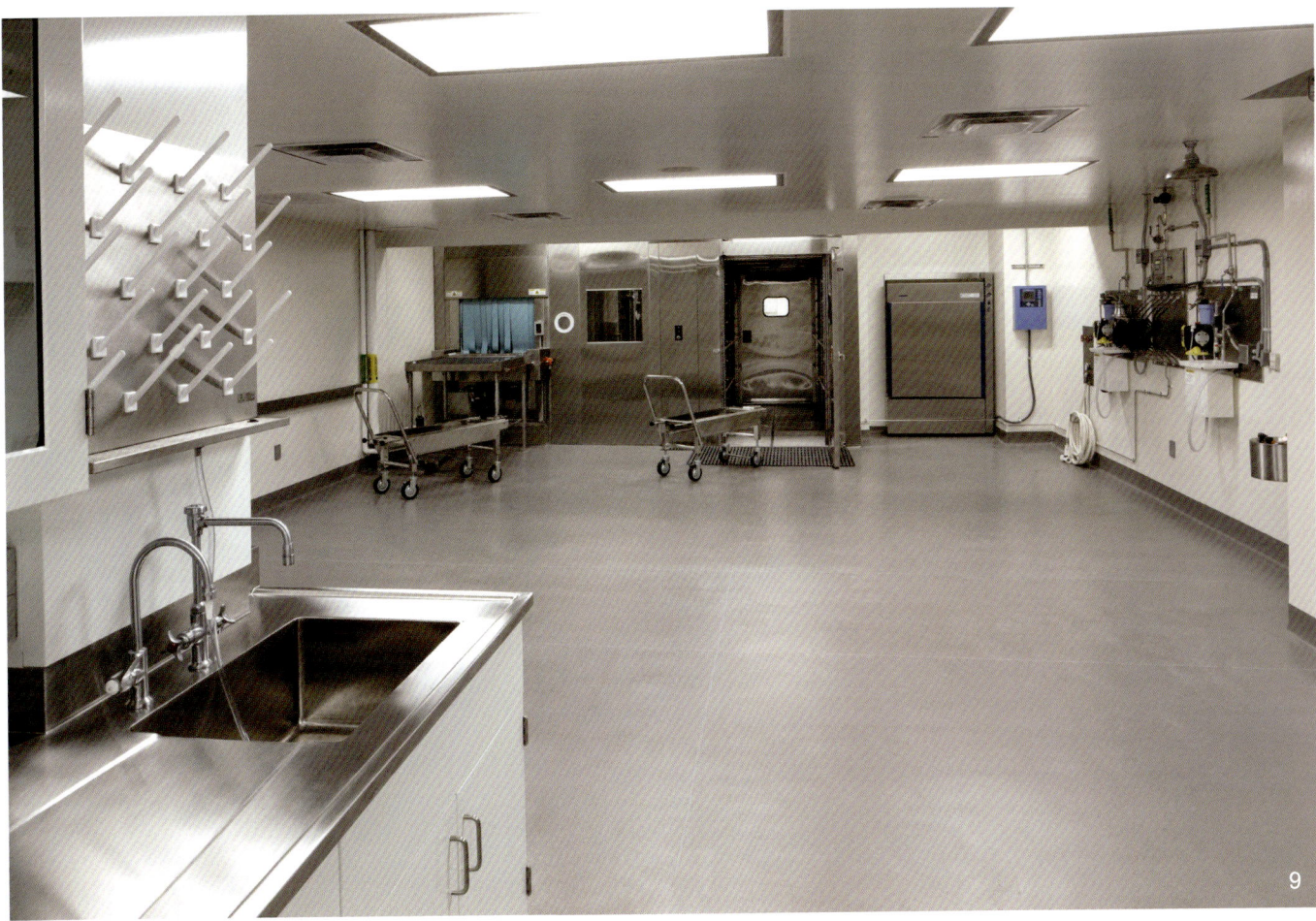

9

1. The long, narrow building is designed to embrace interdisciplinary and collaborative studies in life sciences. It contains conference rooms, lounges, and other common spaces meant to encourage interaction.

2. The architects created a large overhang as a support for solar hot-water generation. The expansive, perforated-steel roof overhang allows filtered light to penetrate in the winter, while also providing shade to the entire south façade in the summer.

3. The two-storey base façade is made from light buff limestone similar in hue to the granite on the signature University of Texas Tower, which is visible from the new facility. The middle layer combines red-brown brick matching adjacent buildings with glass.

4. The main entrance is a two-storey atrium at the southeast corner that opens onto a tree-covered plaza, which is a gathering place for students, staff, and pedestrians.

5. Two secondary entrances are now marked by a recessed, two-storey porch with informal seating, offering a welcome respite from the hot Texas sun.

6. To lessen the amount of ductwork needed, wet-lab facilities are mainly clustered on the two top floors near the rooftop fans so the bulk of ductwork only had to go down through two floors, rather than all six.

7. The labs needed frequent room-air changes—eight per hour in standard laboratories, but between 25 and 45 air changes per hour in organic chemistry labs.

8. In terms of waste disposal, most hazardous wet-lab effluents are detoured into onsite holding tanks, while the remainder is heavily diluted with water to exit through normal sewage pipes.

9. Conference room

TIPS:

1. Texas
The large size of Texas and its location at the intersection of multiple climate zones gives the state highly variable weather. The Panhandle of the state has colder winters than North Texas, while the Gulf Coast has mild winters. Texas has wide variations in precipitation patterns. El Paso, on the western end of the state, averages 8.7 inches (220 mm) of annual rainfall, while parts of southeast Texas average as much as 64 inches (1,600 mm) per year. Dallas in the North Central region averages a more moderate 37 inches (940 mm) per year.

2. ASHRAE 90.1
ASHRAE 90.1 is a standard that provides minimum requirements for energy efficient designs for buildings except for low-rise residential buildings.

3. Labs21
Labs21 is a system for energy-efficient laboratory environmental performance that encourages laboratory owners and operators to make capital investments based on life-cycle cost savings, to pursue advanced HVAC systems, to recover waste heat, and to incorporate renewable energy sources.

Ventilation Metrics

1. Minimum required ventilation rate

Ventilation dominates energy use in most laboratories, especially chemical and biological laboratories. One of the key drivers of ventilation energy use is the minimum ventilation rate required for health and safety. The only exceptions to this are laboratories where the air-change rates are driven by thermal loads (and hence always exceed minimum ventilation rates for health and safety) or where very high fume hood density, typically greater than 1 square foot of hood work surface per 25 gross square feet of laboratory, drives the minimum flow. The purpose of benchmarking minimum ventilation rates is to explore opportunities for optimisation. Specifically, optimisation in this context means reducing air-change rates while maintaining or improving safety. Air-change rates should be benchmarked with two metrics:

Air changes per hour (ACH): This is the most commonly used metric. Various standards and guidelines indicate that this can vary between 4 and 12, which is a very large range. Table 2 shows the range of values listed in various standards. Values higher than 6 ACH (when occupied) and 4 ACH (unoccupied) should be explicitly justified as being required for health and safety.

CFM/sf: Some laboratory professionals believe that this is a more appropriate metric, given that laboratory hazards are more related to floor area than volume, i.e. a laboratory with a high ceiling does not necessarily require more ventilation. The International Building Code (2003) requires a rate of 1 cfm/sf for H-5 hazard environments.

2. Hood density

Fume hoods are prodigious consumers of energy and lab planners should work with owners to carefully avoid installing more and larger hoods than are necessary for programmatic requirements. Specifically, fume hoods should not be used for purposes that can be effectively met with lower-energy alternatives such as snorkels, balance hoods, and chemical storage cabinets. It is recommended that fume-hood density should be benchmarked with other labs that have similar programmatic requirements.

3. Fume hood sash management

Once the number and size of fume hoods has been optimised, the next major opportunity is to reduce fume hood energy use by reducing airflow through low-volume fume hoods and VAV hoods with effective sash management (a major retro-commissioning opportunity).

While there are no commonly used metrics for sash management, we suggest using fume hood airflow management ratio, defined as the ratio of the average flow to the minimum flow. Minimum flow is the flow through the fume hood when the sash is closed. For a typical 6-ft fume hood, this is usually about 300 cfm (which corresponds to the NFPA-45 mandated minimum of 25 cfm/sf of work surface area). A typical 6-ft fume hood with an 18" sash-stop operates at about 900 cfm. Therefore, if the sash were never closed, the airflow management ratio would be 3. If the sash were closed 50 percent of the time, the ratio would be 2.

4. Ventilation airflow efficiency

Ventilation airflow efficiency is typically the most significant way that HVAC design engineers can influence overall lab efficiency. There are two key related metrics:

Pressure drop (in. w.g.): Each component in the supply and exhaust system can be optimised for low pressure drop.

Ventilation system W/cfm: This metric is defined as the total power of supply and exhaust fans divided by the total cfm of supply and exhaust fans. It provides an overall measure of how efficiently air is moved through the laboratory, from inlet to exhaust, and takes into account low pressure drop design as well as fan system efficiency (motors, belts, drives). The figure below shows the range of ventilation system efficiency at peak loads for various laboratories in the Labs21 benchmarking database. There is a wide range of efficiencies, from 0.3 W/cfm to 1.9 W/cfm. The fan power limitations specified in ASHRAE 90.1 2004 provide an additional benchmark. (See Diagram 1)

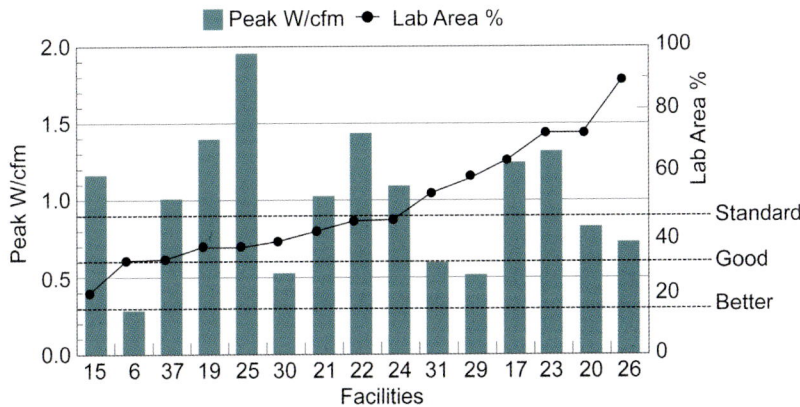

Diagram 1. Ventilation system efficiency at peak conditions for various laboratory facilities in the Labs21 energy benchmarking database. The benchmarks for standard, good, and better practice are based on the Labs21 Best Practice Guide on Low Pressure Drop Design for Laboratories .

Cooling and Heating Metrics – Special Considerations for Labs

1. Temperature and humidity set points

Temperature and humidity set points in laboratory spaces are driven by human comfort and laboratory function (experimentation/equipment requirements). Laboratory users and planners sometimes call for tight tolerances based on laboratory function, without evaluating whether these are actually required. Tight tolerances can increase energy use due to reheat and humidification. It is recommended that tolerances tighter than those required for human comfort (e.g., based on ASHRAE Standard 55 ~12), be carefully evaluated and explicitly justified.

At the Global Ecology Centre at Stanford, equipment requiring tight tolerances (70F+/- 1F) was grouped into a dedicated area so that other areas of the lab could be controlled to wider tolerances (73F+/- 5F) with some rarely accessed freezers and growth chambers actually relocated to a minimally conditioned adjacent structure controlled to 55F ~ 95F.

2. Heating and cooling system efficiency

The key metrics and benchmarks to evaluate the efficiency of chiller and boiler systems in labs are no different than those typically used in other commercial buildings. These include chiller plant efficiency (kW/ton), cooling load (tons/gsf), boiler efficiency (%), pumping efficiency (hp/gpm), etc. Since these are well-documented elsewhere, they are not discussed here and the reader is referred to other publications, such as ASHRAE Standard 90.1. However, two additional metrics have special impact on lab efficiency, and bear further discussion.

Chiller system minimum-turndown ratio: Laboratory systems are often oversized due to reliability/redundancy requirements, over-estimated process loads, or other factors. Even when systems are "right-sized", there are many hours when loads are much lower than peak. Therefore, chiller systems in labs should be designed for low minimum-turndown ratios, defined as the ratio of minimum load (with continuous compressor operation without hot gas bypass or other false loading methods) to design load. Standard practice would be about 20 percent. Good and better practice benchmarks would be 10 percent and 5 percent respectively. In the Molecular Foundry at Lawrence Berkeley National Laboratory (LBNL), the chiller system is capable of a 5 percent turndown ratio. In labs with tight humidity control, even lower ratios are warranted, unless alternative dehumidification strategies are adopted.

Reheat energy-use factor: Reheat energy use can be Reheat energy-use factor: Reheat energy use can be and humidity requirements, wide variation in loads served by a given air handling system 13, or poorly calibrated controls. While there is no well-established metric for assessing reheat energy use, we suggest a metric such as reheat energy-use factor, defined as the ratio of the reheat energy use to the total space heating energy use. The best practice benchmark for this would be 0 percent (i.e. complete elimination of reheat energy use for temperature control).

Plug Load Metrics

Equipment loads in laboratories are frequently overestimated because designers often use estimates based on "nameplate" data, and design assumptions of high demand. This results in oversized HVAC systems, increased initial construction costs, and increased energy use due to inefficiencies at low part-load operation 16. The following related metrics can be used to assess and compare design and measured plug loads:

Laboratory design plug load W/sf: The values may vary across lab spaces in a given building. Note that the assumption for electrical system design is usually higher than that for HVAC system design.

Laboratory actual (measured) plug load W/sf: This is obtained by taking continuous measurements at the panel serving laboratory plug loads. For HVAC system design, it is more appropriate to consider the maximum of the 15-minute interval averages (rather than maximum instantaneous load), since HVAC systems typically do not react to the instantaneous loads. For a building currently in design, it is recommended that measurements be taken in a comparable laboratory and those data be used for sizing.

The figure below compares the measured peak loads (maximum instantaneous and maximum 15-minute interval average) to the design loads for various laboratory spaces in a building at the University of California, Davis. While the sizing ratio is driven by context-specific factors such as reliability and flexibility, it is recommended that sizing factors greater than 2 be carefully evaluated and justified. (See Diagram 2)

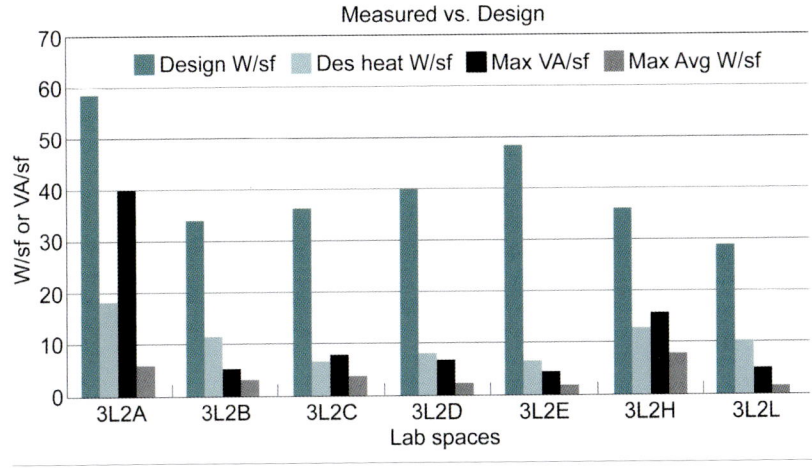

Diagram2.Comparison of design loads and measured plug loads in various laboratory spaces at the University of California, Davis. Measurements were taken over a 2-week period while labs were fully occupied. Des W/sf is the peak plug load assumption for electrical design. Des heat W/sf is the peak plug load assumption for HVAC design. Max VA/sf is the measured peak (instantaneous) apparent power. Max Avg W/sf is the maximum of the 15-minute averages.

Lighting Metrics

The key metrics and benchmarks to evaluate the efficiency of lighting systems in laboratories are not fundamentally different than those typically used in other commercial buildings. These include daylight factors, illuminance levels, lamp and ballast efficacy, lighting power density, etc. There are two key metrics for which the benchmarks in laboratories are different from other commercial buildings:

Task illuminance in laboratory spaces (fc): The 9th edition of the IESNA Handbook 17 has revised its illuminance recommendations for laboratories downward from the previous edition. The current recommendations are:
• Specimen collecting: 50 fc (horizontal), 10 fc (vertical)
• Science laboratory: 50 fc (horizontal), 30 fc (vertical)

Values higher than 50 fc should be carefully reviewed and justified by special functional requirements and should be restricted to the areas where the task is being performed. Furthermore, it is important to recognise that illuminance in and of itself is not an adequate measure of visual acuity, which is a function of several other factors, such as contrast ratios, colour rendition, etc.

Installed lighting power density (W/nsf): This refers to the lighting power density in the laboratory spaces.

ASHRAE 90.1-2004 allows a maximum of 1.4 W/sf. The California Title 24 energy code allows a maximum of 1.3 W/sf. At the Tahoe Centre for Environmental Studies, the laboratory spaces were designed to 0.80 W/sf.

How to Specify and Track Metrics – Process Considerations

The following are some key process considerations to specify and track metrics during design, delivery, and operation of laboratory buildings:

(1) **Identify metrics and set targets with stakeholder team.** Metrics and targets are, in effect, key performance indicators for the quality of design and operation, and therefore should have the buy-in of all the key stakeholders (owners, designers, and operators). This could be done at project conception, and then refined during the early stages of the project. In the design for a new laboratory at LBNL, for example, a goal-setting meeting was held prior to conceptual design, in which the designers and owners considered a wide range of metrics, selected key metrics, and set targets for them. The list of metrics in Appendix A could be used as a template for identifying metrics and setting targets.

(2) Incorporate key metrics and targets in programming documents. Designers and operators are much more likely to ensure that targets are met if they are officially incorporated into the programming documents.

(3) Identify individual(s) responsible for tracking metrics. Ideally, the commissioning authority would have overall responsibility, since metrics are integral to the performance tracking and assurance process. However, various design professionals may have responsibility for computing individual metrics and providing these to the commissioning authority (e.g. lab planner for hoods/nsf, HVAC engineer for W/cfm, etc.)

(4) Determine process and format for tracking and documenting metrics. The Labs21 Design Intent Tool can be used to track metrics and generate formatted reports in a consistent manner over the course of a project.

References

1. Research and Technology Buildings – A Design Manual, published by Birkhäuser
2. "High Performance Laboratories " Pacific Gas and Electric Company (PG&E)
3. Lawrence Berkeley National Laboratory, Metrics and BenchMarks for energy efficiency in LaBoratories
4. Laboratories For the 21st Century: Best Practices, Daylighting In Laboratories
5. Laboratories For the 21st Century: Best Practice Guide, Efficient Electric Lighting in Laboratories
6. Laboratories for the 21st Century: Best Practices, Energy Recovery in Laboratory Facilities
7. Laboratores for the 21st Century: Best Practces, Water Efficiency Guide for Laboratories
8. Trends in Lab Design, by Daniel Watch, Available at http://www.wbdg.org/resources/labtrends.php?r=research
9. Research Laboratory, by Daniel Watch and Deepa Tolat, Available at http://www.wbdg.org/design/research_lab.php
10. Richards Medical Research Laboratories Building, Available at http://www.workshopoftheworld.com/west_phila/richards.html
11. http://www.workshopoftheworld.com/west_phila/richards.html

INDEX

JOSÉ JUAN BARBA Architect
http://www.josejuanbarba.com/portfolio/project_2/info.html
Metalocus. C/ Canarias, 5 - 1 C. 28045 - Madrid, Spain.
Tel: +34 915 396 976
Fax: +34 912 225 056

Kopper Architektur
http://www.kopper-architekten.de/

Lahznimmo Architects
http://lahznimmo.com/
Suite 404 Flourmill Studios, 3 Gladstone Street, Newtown NSW
2042 Australia
Tel: +02 9550 5200
Fax: +02 9550 5233

Legat Architects, Inc.
http://www.legat.com/
651 W. Washington Blvd., Suite 1, Chicago, IL 60661
Tel: +312.258.9595
Fax: +312.258.1555

Lyons
http://www.lyonsarch.com.au/
Level 3, 246 Bourke Street, Melbourne, Victoria 3000, Australia
Tel: +61 3 9600 2818
Fax: +61 3 9600 2819

Marlene Imirzian & Associates LLC
http://www.imirzian-architects.com/
8906 North Central Avenue, Phoenix, AZ 85020
Tel: +602 943 5279

NAC Architecture
http://www.nacarchitecture.com/
2025 First Avenue Suite 300, Seattle, WA 98121-3131
Tel: +206 441 4522
Fax: +206 441 7917

Planho Consultores
http://planho.com/
C/ Arjona 10 Esc 3 1°C, 41001 Sevilla
Tel: +954 50 25 10

Rudy Uytenhaak Architectenbureau BV
http://www.uytenhaak.nl/
Rijnsburgstraat 9, 1059 AT Amsterdam, The Netherlands
Tel: +31 0 20 305 77 77
Fax: +31 0 20 305 77 78

Spillman Farmer Architects
http://www.spillmanfarmer.com/
1720 Spillman Drive Suite 200
Bethlehem, Pennsylvania 18015
Tel: +610.865.2621
Fax: +610.865.3236

Stanton Williams Architects
http://www.stantonwilliams.com/ie6/
36 Graham Street, London N1 8GJ, UK
Tel: +44 (0)20 7880 6400

Styria Arhitektura
http://www.styria.si/
Cankarjeva ulica 6e, 2000 Maribor, Slovenia
Tel: +386 2 228 29 88
Fax: +386 2 228 29 89

Taniguchi Architects
http://www.taniguchi-arch.com/
1609 West Sixth Street, Austin, Texas 78703
Tel: +512 474 7079
Fax: +512 474 7579

Wilson Architects
http://www.wilsonarchitects.com.au/
564 Boundary Street
Spring Hill Qld 4000
Tel: +07 3831 2755

© 2013 by Design Media Publishing Limited
This edition published in October 2013

Design Media Publishing Limited
20/F Manulife Tower
169 Electric Rd, North Point
Hong Kong
Tel: 00852-28672587
Fax: 00852-25050411
E-mail: suisusie@gmail.com
www.designmediahk.com

Editing: Neil Appleton
Editorial Assistant: Helen Liu
Proofreading: Maggie Wang
Design/Layout: ZHOU Jie

ISBN 978-988-15664-7-8

Printed in China